Sustainable Development

Anne E. Egelston

Sustainable Development

A History

Springer

Dr. Anne E. Egelston
Government Department
Lone Star College-Montgomery
Conroe, TX 77384
USA

ISBN 978-94-007-4877-4 ISBN 978-94-007-4878-1 (eBook)
DOI 10.1007/978-94-007-4878-1
Springer Dordrecht Heidelberg New York London

Library of Congress Control Number: 2012948992

Printed on acid-free paper

Springer is part of Springer Science+Business Media (www.springer.com)

Acknowledgments

My dream of publishing this book would not have been possible without the generous love, time, and support of my parents, Tom and Carol Egelston. This book began life as a dissertation in 2005, but languished on the shelf for 4 years after I graduated and I am astounded that it was ever published at all. I continue to owe debts of gratitude to my advisors at Rutgers University – Newark.

I also want to thank and acknowledge all of the time spent on the phone to my brother, John and his wife, Wendy, whose expertise in editing and proofreading extended far and above any reasonable request of a sister-in-law. This is especially true when the phone call travels half way around the world.

To my grandfather, Wayne Lewallen, who insists that he needs a footnote.

To Eleanor, John, Dorris, and Laura, who weren't able to see this to completion.

To Princess, Cowboy, Pumpkin, Claire Bear, and Stinky Pants, you're stuck with me, I'm afraid.

Contents

List of Abbreviations

BINGO	Business and Industry Non-Governmental Organization
CSD	Commission on Sustainable Development
ECOSOC	Economic and Social Council
ELCI	Environment Liaison Centre International
ELB	Environment Liaison Board
ENGO	Environmental Non-Governmental Organization
GNP	Gross National Product
INASEN	International Assembly of Non-Governmental Organizations Concerned with the Environment
ICC	International Chamber of Commerce
ICSU	International Council for Science
IIED	International Institute for Environment and Development
IIEA	International Institute of Environmental Affairs
IUCN	International Union for the Conservation of Nature and Natural Resources
IWC	International Whaling Commission
NIEO	New International Economic Order
NGOs	Non-Governmental Organizations
OECD	Organization for Economic Co-operation and Development
PrepComs	Preparatory Committees
SCOPE	Special Committee for Problems on the Environment
TNC	Transnational Corporation
UN	United Nations
UNCED	United Nations Conference on Environment and Development
UNCHE	United Nations Conference on the Human Environment
UNCTAD	United Nations Conference on Trade and Development
UNESCO	United Nations Educational, Scientific, and Cultural Organization
UNEP	United Nations Environment Program
WCED	World Commission on Environment and Development
WSSD	World Summit on Sustainable Development
WTO	World Trade Organization

Chapter 1
An Invitation to Explore the Role of NGOs

Abstract The chapter invites the reader to investigate how non-governmental organizations influenced the sustainable development discourse. Nongovernmental organizations' influence may not have been known and predictable, but their influence was significant. The traditions established at Stockholm largely remain intact as diplomats prepare for the upcoming Rio+20 United Nations Conference on Sustainable Development in 2012.

The chapter also provides information on the research design and methodology as well as introducing the chapters in the remainder of the book.

Keywords Sustainable development discourse • Non-governmental organization influence • Research design/methodology • Environmental diplomacy • United Nations mega-conference

1.1 Environmental Consciousness

Since the beginning of widespread environmental political consciousness, Spaceship Earth stood as a beautiful bright blue and white marble against the black background of outer space. This image brought back to earth by the Apollo astronauts, more so than anything else, encouraged grassroots organizations located in the United States and elsewhere in the world to give voice to their shared, damaged environment. Other events such as the publication of Rachel Carson's book *Silent Spring* in 1962 and the first Earth Day in 1970 helped to solidify these grassroots organizations into national and international coalitions.

Yet a mere 35 years after these events, the future for the environmental movement does not seem quite so picturesque. Regular reports of continuing environmental pollution appear on the nightly news. Climate change, contaminated water, nuclear waste disposal and poor air quality have become all too familiar problems. First generation environmental problems were easily identified and lent themselves to technical solutions such as clean air and clean water. These problems gave way to

A.E. Egelston, *Sustainable Development*, DOI 10.1007/978-94-007-4878-1_1,

second generation issues that were not so-easily solvable such as non-point source runoff and automobile emissions. Third generation problems, ones that are global in both cause and effect, such as biodiversity loss, desertification, and climate change have proven to be much more intractable. Negotiations on these issues have lasted decades or longer with no end in sight for either the negotiations or the activities that have caused such menacing environmental risks.

Further questions about who governs, the level of environmental protection codified by regulations, and what constitutes appropriate responses from regulated entities continue to consume significant resources. The conceptualization of sustainable development, a major environmental discourse, continues to be problematic as states and transnational corporations (TNCs) continue to spend billions of dollars reacting to incessant demands for better environmental performance. Despite the fact that significant worldwide investments are made in sustainable development practices yearly, little is known about the early political history of sustainable development, its supports, detractors, goals, and preferred outcomes.

This book characterizes sustainable development as a discourse, in line with Dryzek's (2005:9) exposition that a "discourse is a shared way of apprehending the world." Discourses rely upon language to socially construct new relationships or meanings, or to alter existing relationships or meanings. Consequently, sustainable development can be viewed as a holistic approach to the relationship between man and the environment. Yet the sustainable development discourse does not have one unique set of meanings. Various actors emphasize differing facets of sustainable development depending on their needs. Business and industry groups focus on waste elimination and health and safety improvements. Minority groups focus on equal access to environmental resources. Simple living organizations focus on reduced consumption patterns.

This book answers inquiries into the political history of sustainable development by looking at the evolution of sustainable development and the non-governmental organizations (NGOs) that promoted this discourse prior to the formation of the World Commission on Environment and Development (WCED), also known as the Brundtland Commission. Consequently, this book is also about how NGOs, in particular environmental NGOs (ENGOs), influenced international organizations such as the United Nations Environment Program (UNEP).

These non-state actors first articulated many important ideas and concepts associated with sustainable development from the earliest moments of the United Nations Conference on the Human Environment (UNCHE) in 1972. However, scholars tend to discuss sustainable development as if it appeared for the first time with *Our Common Future*, a politically driven manifesto that permanently established sustainable development on the UN agenda in 1987. Instead, NGOs established many traditions for future interactions with other actors at Stockholm. They also successfully executed their plans to establish a permanent working relationship with UNEP.

As this book moves towards publication, preparations for Rio+20, the United Nations Conference on Sustainable Development (UNCSD) are well underway. This meeting celebrates its predecessor conference, the United Nations Conference on

Environment and Development (UNCED) also known as the Earth Summit, that heralded sustainable development's meteoric rise to the dominant environmental discourse within international negotiations. Advocates supporting the Rio+20 conference hope for a break-through moment that will further elevate sustainable development from a discourse to an international regime, complete with binding international treaties and enforcement mechanisms. Like the conferences before it, Organizers for Rio+20 expect wide-spread public participation from NGOs both within, and outside, the formal conference halls.

That having been said, very few people expect diplomats to create any meaningful treaties at this meeting. Instead, the objective of the Rio+20 conference is to reinvigorate political energy, raise media attention and stimulate further voluntary actions from all levels of government implementing sustainable development. In the midst of one of the worst periods of economic uncertainty sometimes referred to as the Great Recession and growing numbers of states imposing austerity measures, it seems highly unlikely that funding for any significant institutional reforms or increased financial aid will emerge as a result of this upcoming conference.

That is not to say the implementation of sustainable development has halted. It has not. If anything, awareness of the values enshrined within sustainable development has increased since its inception. This trend is likely to continue in the future, despite the economic downturn. NGOs will continue to use their rational moral authority to influence key actors, both within, and outside, the United Nations (UN).

1.2 NGOs Political Influence on Sustainable Development

The focus of this book is upon a specific type of NGOs, environmental organizations recognized nominally by the UN as having special expertise on the environment and/or upon human interactions with the environment. Some prominent NGOs within this category include Friends of the Earth (FoE), Greenpeace, International Union for Conservation of Nature (IUCN), and World Wildlife Federation (WWF). Other NGOs also participate at the UN such as the International Chamber of Commerce (ICC) or Red Cross, but these organizations self-identify in other subcategories of NGOs such as business and industry NGOs (BINGOs) or humanitarian NGOs.

Academic literature documenting NGOs' influence in the past has focused on NGOs' activities during specific environmental conferences and their associated regimes including biodiversity (Arts 1998), climate change (Ahrens 2002; Betsill 2000), deforestation (Bramble and Porter 1992), desertification (Corell 1999), the Arctic (Young 1992, 1998; Young and Osherenko 1993), marine dumping (Stairs and Taylor 1992), and ozone depletion (Litfin 1994). Additionally, scholars have studied NGOs' influence with respect to the final negotiating text at the UN conferences that form the main negotiating backdrop for sustainable development – at Stockholm (Feraru 1974), Rio de Janeiro (Chatterjee and Finger 1994; Finger

1994) and Johannesburg (Gutman 2003; Mwangi 2005). This book expands the focus on NGOs' influence from a specific negotiating session to the full time frame of diplomatic negotiations around sustainable development. Including "time" as a key variant immediately complicates efforts to isolate instances of NGOs' influence. Time increases the complexity of analysis by adding more data to the case study. This addition compounds the potential number of activities that could create influence. On a more positive note, widening the time frame allows scholars to investigate more facets of NGOs' activities as there is no longer a need to utilize only changes in treaty text as a proxy for NGOs' influence.

This book begins from the premise that NGOs should be present at the negotiating table and meaningfully contribute to the ordering of international life. Further, NGOs are a useful and necessary political actor with unique roles that states are unwilling or unable to fulfill. That having been said, not all NGO influence is intentional; the future is inherently unknown. One can suspect that acting in manner A will achieve outcome B. On the other hand, NGOs may in fact achieve outcome C and that outcome may be more or less important than the original goal and totally unexpected. The outcome may not be intentional, but the impact may well be significant.

This book primarily takes a constructivist approach to this research. It reviews the time period when the majority of the international institutions, patterns of engagement, and cultural context of global environmental politics were formed. The freedoms and flexibilities, as well as the constraints and prohibitions upon various elements of the current institutional arrangements were proposed, accepted, and formalized in writing during this time period. That is not to say that all of these things remain set in stone today, but rather to acknowledge that the creation of the sustainable development discourse is situated in time and in place. Thus, this research is not intended to produce general theory, but rather to explain a specific time and place with the understanding that future times and places might be sufficiently similar to the case study and that this specific theory might be utilized again.

1.3 Research Design

This project was initially conceived to examine the influence of business groups at three UN environmental conferences – UNCHE (Stockholm, 1972), the UNCED (Rio de Janeiro, 1992) and the World Summit on Sustainable Development (WSSD) (Johannesburg, 2002). It quickly became apparent that the kinds of information needed to perform this research were not readily available due to the loss of original "unofficial" documentation that so often occurs with the passing of time. Documents, particularly from the 1972 Stockholm Conference were preserved on an *ad hoc* basis, with a notable bias towards pro-environmental groups. Documents about business non-governmental organizations were generally unavailable and this skewed the evidence towards environmental organizations that generally kept better records. Additionally, existing literature on NGOs, contains a great deal of material on the influence of NGOs on states at a specific point in time. However, scholars had conducted very little research concerning the impact of NGOs on shaping the sustainable development discourse. This book, then centers around questions relating to the conceptualization of sustainable development, its proponents, and the tactics used to permanently alter global environmental politics.

The research phase of this project began by relating the key events leading to the creation of the WCED. Very little evidence existed suggesting significant state investment into the conceptualization of sustainable development. Consequently, the research shifted to an examination of the role of non-state actors in promoting sustainable development.

The research utilizes a single qualitative case study to examine the role of NGOs in promoting and implementing sustainable development from the beginning of the UNCHE in 1968 to the establishment of the WCED in 1983. Many activities promoting the discourse underlying sustainable development occurred during this time frame, although the events were neither highly publicized nor recognized as trailblazing when they occurred.

Sustainable development entails a radical change in the current socio-economic system; because of the unusual depth of change expected, it has been slow to form when compared to other simpler regimes.[1] It therefore provides a unique case study, not only because of the unusual lead-time, but also because of the numerous interconnections between sustainable development and the more nebulous governance provided by NGOs. Thus, sustainable development has both a substantial negotiating history, and, due to its slow pace of formation, an opportunity to substantially affect the future of the regime, should it come into existence. Accordingly, this research provides a timely contribution not only to the scholarly literature on the environment, but also potentially for practitioners in the hallways of the UN. Sustainable development also retains the narrow focus of a single, albeit large, case study while retaining the "real world" complexity of actors and preferred outcomes.

1.4 Methodology

This research consists of a single case study, the case of sustainable development. In this instance, a single case study is an appropriate selection because of its uniqueness in that it represents a new subset of cases studies (Yin 2003:40). Sustainable development continues to evolve during the course of the last 30 years due to its complex nature. The case study also served as a pioneer of sorts for NGO participation in UN meetings and it also demonstrated the complexity of underlying

[1]The Law of the Seas took 12 years to enter into force. While climate change discussions began in 1988, binding commitments to reduce greenhouse gases did not enter into force until 2005.

causes and potential solutions to intractable environmental problems. A case study is a description of an event in such a way as to give primacy to observation, reconstruction, and analysis (Zonabend 1992).

Case studies rely on the manner in which the case is constructed in order to satisfy methodological rigor (Yin 2003). Case studies are ideal for research when a better understanding of the dynamics behind specific actions is sought (Eisenhardt 1989:534). Moreover, Platt (1988:20) states, "openness to surprise and availability for multiple purposes is a real strength." As my research question focuses on how NGO influence promoted the sustainable development discourse, a case study is an appropriate methodological selection for this inquiry. Yin (2003:3) subdivides case studies into three types – exploratory case studies, descriptive case studies, and explanatory case studies. This case study is descriptive in the sense that a great deal of historical research has been necessary to construct the appropriate framework for investigation.

The research material for this case study consisted of original UN documentation, media reviews, NGO publications and other scholarly works. Triangulation of sources and techniques (Corell and Betsill 2001), was utilized in order to control for source bias. Stake (1995) identifies at least six sources of evidence in case study: documents, archival records, interviews, direct observation, participant-observation, and physical artifacts. Evidence that does not lead to triangulation of sources has been disregarded as unreliable.

Background information came from archival research conducted at the Environment Information Center at Harvard University, particularly the Maurice F. Strong Collection, and the Barbara Ward (Baroness Jackson) papers archived at the Special Collections Department of Georgetown University. As this case study is primarily historical in nature, a thorough review of both primary and secondary literature provided the bulk of research. These sources allow the reconstruction of the events that took place at these historic conferences. Final negotiating text, while important, provides a limited point of access towards this kind of reconstruction. Therefore, I looked for "unofficial" documents such as position papers, written reports, UN Conference Secretariat press releases, and newspaper articles published concomitantly with the conferences to provide the majority of materials for this book.

In recent years, UN conferences are notable for the amount of documentation produced by advocacy groups. The first environmental conference probably established this pattern, as it did with so many other traditions. That is not to say, that all of these records are well-preserved. Unfortunately, some of the supporting documents of these conferences and unremarkable reports from this time-period have been lost, as have the diplomats and observers who struggled to keep environmental affairs on the map during a time when environmental policies suffered from both a lack of political will and a lack of funding.[2] Strength of

[2]Selin and Linnér (2003) provide an outline of events between Stockholm and Rio. However, they do not attempt to analyze the changes in rhetoric that signify an underlying shift in norm development, which in turn provides the organizational frame for a significant portion of this case study.

evidence varies by event and special care has been taken not to overstate impacts based upon available evidence.

Participant observation at other UN conferences was helpful in the initial phases of this research as it gave insight into UN procedures and processes. International negotiating sessions have their own rules of conduct and standards of behavior that are best understood through personal experience and observation. Meetings attended included the Commission on Sustainable Development (CSD) in 2000, as part of the ICC and multiple meetings of the UN Framework Convention on Climate Change in 1999, 2000, and 2003 as part of the Emission Marketing Association. While none of these meetings generated observations for analysis, per se, attendance as part of a BINGO have shaped my thoughts and insights in understanding not only the formal processes of diplomacy, but also the ebb and flow of side-bar events.

Secondary resources such as previously published works provided useful information as they contained details about the conferences, particularly from the many NGO participants in the form of documenting personal experience (Artin 1973; Gutman 2003; Stone 1973; Viña et al. 2003). These works have proved invaluable in reconstructing side-bar events, hallway meetings, and informal bargaining sessions.

Lastly, conferences generally have one or more independent daily newspapers in circulation for the duration of the event. Like the participant observations, these articles provide evidence of the important issues and context that surround the meetings. In certain instances, the daily newspapers provide important insights on issues that are conspicuous by their absence from the negotiating agenda.

1.5 Remainder of the Book

The second chapter of this book provides background information for the reader. Global environmental politics continues to involve complex negotiations with a multitude of actors, agents, agenda, norms, and goals. This chapter sets the state for a detailed look at NGOs and their relationship with the UN, including UNEP.

The third chapter of the book establishes the model used for determining NGO influence during the conceptualization of the sustainable development discourse. The first half of the book reviews various strains of thought within international relations theory about the relationships between states and NGOs. It details NGOs role as actors within global environmental governance who may seek to influence states, international organizations, NGOs, or other members of global civil society. The second half of this chapter examines NGOs' influence with respect to international environmental negotiations and extends this theory into a different political arena, namely the creation of the sustainable development discourse. The chapter also includes theoretical conclusions about the conditions that might allow NGO to have more (or less) influence during this case study.

The fourth chapter provides background information on the institutional history of sustainable development, beginning in 1968 with the UN processes leading up to the Stockholm conference. It continues by filling in the oft-dismissed history between the formation of UNEP and the WCED that produced *Our Common Future*, more commonly referred to as the Brundtland Commission and the Brundtland Report, respectively. The chapter highlights the political process that turned the ecodevelopment thesis of the Stockholm conference into sustainable development.

The fifth chapter presents the analysis of the case study. The main finding is that NGOs role, function and political influence began with the Stockholm conference negotiating process in 1970 and continues unaltered and uninterrupted throughout the remainder of the creation of sustainable development. NGOs have significant political influence and have kept sustainable development "alive" at a time when few, if any, states were interested in the concept. Further, NGO influence was enabled by both time and place; in short, state's indifference towards UNEP, coupled with UNEP's desire to accomplish its tasks, created a political problem that NGO resources helped resolve.

The sixth, and final chapter, will draw conclusions about this research with a particular emphasis on the current and future roles of NGOs at the UN, as well as the status of sustainable development. It also speculates on one potential future of sustainable development and assesses the likelihood of this prediction. Finally, the book concludes by examining the limitations of this theoretical model and outlining avenues for future research.

References

Ahrens, L. (2002). *Explaining the role and political influence of non-governmental organizations in international decision-making process*. Thesis, Rutgers, The State University of New Jersey.

Artin, T. (1973). *Earth talk independent voices on the environment*. New York: Grossman Publishers.

Arts, B. (1998). *The political influence of global NGOs: Case studies on the climate and biodiversity conventions*. Utrecht: International Books.

Betsill, M. (2000). *Greens in the greenhouse: Environmental NGOs, norms and the politics of global climate change*. Dissertation, University of Colorado.

Bramble, B. J., & Porter, G. (1992). Non-governmental organizations and the making of US international environmental policy. In A. Hurrell & B. Kingsbury (Eds.), *The international politics of the environment: Actors, interests and institutions* (pp. 313–353). Oxford: Clarendon.

Chatterjee, P., & Finger, M. (1994). *The earth brokers. Power, politics and world development*. London: Routledge.

Corell, E. (1999). *The negotiable desert: Expert knowledge in the negotiations of the convention to combat desertification*. Dissertation. Linkoping University.

Corell, E., & Betsill, M. (2001). A comparative look at NGO influence in international environmental negotiations: Desertification and climate change. *Global Environmental Politics, 1*(4), 86–107.

Dryzek, J. S. (2005). *The politics of the earth: Environmental discourses* (2nd ed.). Oxford: Oxford University Press.

Eisenhardt, K. M. (1989). Building theories from case study research. *Academy of Management Review, 14*(4), 532–550.

Feraru, A. T. (1974). Transnational political interests and the global environment. *International Organization, 28*(1), 31–60.

Finger, M. (1994). Environmental NGOs in the UNCED process. In T. Princen & M. Finger (Eds.), *Environmental NGOs in World Politics Linking the Local and the Global*. New York: Routledge.

Gutman, P. (2003). What did WSSD accomplish? An NGO perspective. *Environment, 45*(2), 20–28.

Litfin, K. (1994). *Ozone discourses*. New York: Columbia University Press.

Mwangi, W. (2005). Three decades of NGO activism in international environmental negotiations. In A. Churie, G. Sjöstedt, & E. Corell (Eds.), *Global challenges furthering the multilateral process for sustainable development* (pp. 171–193). Sheffield: Greenleaf Publishing.

Platt, J. (1988). What can case studies do? *Studies in Qualitative Methodology, 1*, 1–23.

Selin, H., & Linnér, B. -O. (2003). *The quest for global sustainability: International efforts on linking environment and development* (Working Paper No. 5). Cambridge, MA: Science, Environment and Development Group, Center for International Development, Harvard University.

Stairs, K., & Taylor, P. (1992). Non-governmental organizations and the legal protection of the oceans: A case study. In A. Hurrell & B. Kingsburg (Eds.), *The international politics of the environment* (pp. 110–141). Oxford: Oxford University Press.

Stake, R. (1995). *The art of case research*. Thousand Oaks: Sage.

Stone, P. B. (1973). *Did we save the earth at Stockholm?* London: Earth Island Limited.

Viña, A. G. M., Hoff, G., & Derose, A. M. (2003). The outcomes of Johannesburg: Assessing the world summit on sustainable development. *SAIS Review: A Journal of International Affairs, 23*(1), 53–70.

Yin, R. (2003). *Case study research: Design and methods* (3rd ed.). Thousand Oaks: Sage.

Young, O. (1992). *Arctic politics: Conflict and cooperation in the circumpolar north*. Hanover: University Press of New England.

Young, O. (1998). *Creating regimes: Arctic accords and international governance*. Ithaca: Cornell University Press.

Young, O., & Osherenko, G. (1993). *Polar politics: Creating international environmental regimes*. Ithaca: Cornell University Press.

Zonabend, F. (1992). The monography in European ethnology. *Current Sociology, 40*(1), 49–60.

Chapter 2
NGOs Herald the Arrival of Sustainability

Abstract The chapter provides background information about the United Nations-nongovernmental organization relationship and how these rules have developed over time. It opens by presenting background information about the United Nations system and gives a brief overview of one of its organs – the United Nations Environment Program. The chapter continues with a review of the United Nations Environmental Conferences focusing on the Stockholm-Rio-Johannesburg Trajectory that highlights major turning points in the ongoing diplomatic negotiations about sustainable development.

The remainder of the chapter introduces the sustainable development discourse with a focus on its importance as a political compromise that allowed Northern developed countries to discuss environmental protection with Southern developing countries. As a result sustainable development remains a controversial topic representing different philosophies ranging from resource conservation to socio-economic equality through environmental justice.

Keywords Non-governmental organizations • UN mega-conferences • Consultative status • Sustainable development • United Nations Conference on Human Environment (UNCHE) • United Nations Conference on Environment and Development (UNCED) • World Summit on Sustainable Development (WSSD)

The UN system came into existence towards the end of World War II as part of a new infrastructure that could be used to resolve conflicts without resorting to violence and therefore could contribute to international cooperation and stability. The structure of the UN changes over time; however, the two main organs likely to be involved within environmental affairs are the General Assembly and the UN Secretariat. The General Assembly consists of the member states, with each state holding one vote. The Secretary-General heads the Administration of the UN and is elected to a 4-year term by the General Assembly. As a result of this arrangement,

A.E. Egelston, *Sustainable Development*, DOI 10.1007/978-94-007-4878-1_2,

the numerically superior Southern developing countries have a natural majority within the General Assembly, but may be unable to raise financial support for their platform.

The General Assembly generally asks the UN Secretariat to undertake a task – in this case, the convening of a conference. This request makes the Secretary General responsible for supporting the international negotiating committee that convenes to discuss the agenda for the conference. In 1968 the General Assembly issued Resolution 2398 (XXIII) asking the Secretary General to convene the UNCHE and this conference became the first major discussion of international environmental policies. Items on the meeting agenda included formal treaties (hard law), as well as statements, principles and traditions (soft law).[1] This conference typically marks the coming of age of international environmental negotiations (Caldwell 1984; Caroll 1988; Hurrell and Kingsbury 1992). Sporadic activity occurred under the auspices of the League of Nations (Charnowitz 1997; White 1951), prior to the UN. Additionally, the United Nations Educational, Scientific, and Cultural Organization (UNESCO) sponsored information exchanges on the environment (Hoggart 1996).[2]

2.1 UNEP

After the first UNCHE, the General Assembly created UNEP as a program reporting to the UN Secretary General rather than as a more powerful specialized agency. Ultimately, the UN General Assembly tasked UNEP with coordinating the other specialized agencies with a small budget and office staff far removed from the central hallways of the UN in New York and Geneva.

The organizational model for UNEP differed quite markedly from the newly created UN Development Program (UNDP). UNDP benefitted from a program-oriented focus, complete with substantial funding in the form of overseas development aid. Instead, UNEP received the unenviable task of coordinating the environmental components within a highly fragmented UN system. UNEP's role is "to promote international cooperation in the field of the environment, and to recommend, as appropriate, policies to this end; [and] to provide general policy guidance for the direction and co-ordination of environmental programs within a United Nations system (UN 1972)." This role severely limits UNEP's capabilities to advance a pro-environmental agenda as UNEP was not intended to be an action oriented advocate, but rather constrained to advising other entities, who were then free to adopt or ignore UNEP's policy guidance.

---

[1] For more information about hard and soft law, see Abbott and Snidal (2000) and Sheldon (2000).

[2] Although some rare instances of treaty negotiations (and treaties, mainly involving the protection of migratory birds) occurred, many of these conferences could be considered as non-political, informational exchange meetings, such as the 1968 UNESCO sponsored Man and the Biosphere Conference.

UNEP is led by an Executive Director, who is appointed by the UN Secretary-General for a 4-year term. To date, UNEP's work has been overseen by five Executive Directors. UNEP takes programmatic direction from the Governing Council, a group of 58 member states. The member states comprising the Governing Council are elected by the General Assembly based upon equitable geographic distribution, ensuring that developing countries controlled UNEP. The Governing Council also holds approval authority for UNEP's budget, and meets every 2 years in Nairobi. The Governing Council assembles in Special Session in the off-year away from Nairobi.

Calls to strengthen UNEP's catalytic role regularly occurs and a lively debate over the necessity of a World Environment Organization echo particularly loudly in advance of the 20th anniversary of the 1992 Earth Summit. At the time of writing, member states and major groups participating in the negotiations leading up to the "Rio+20" conference on June 20–22, 2012 are considering options for strengthening the institutional framework for sustainable development. The options include reform of existing institutions such as UNEP, the CSD, and Economic and Social Council (ECOSOC), as well as creating new institutions such as a World Environment Organization or a new umbrella organization for sustainable development.[3]

In spite of the complex international bureaucracy that comprises the UN machinery, UNEP reports that over 500 multilateral environmental agreements have been brought to fruition (UNEP 2001: ii). Of these multilateral environmental agreements over 300 were completed since 1972, a substantial increase in activity. Most prominent among these treaties are accords to address issues such as climate change, endangered species, atmospheric ozone depletion, wetlands, deforestation, land degradation, protection of the seas, hazardous waste dumping, and biodiversity. Consequently, academic supporters of UNEP (von Moltke 1996; Najam 2001), consider this catalytic role a success.

2.2 The UN Environmental Conferences

Regardless of the formal role and responsibility of UNEP, the UN General Assembly continues to create international negotiating committees to advance environmental affairs. The *ad hoc* UN multilateral mega-conferences, also known as summits, are authorized by a resolution that specifies the arrangements for a preparatory conference. The resolution typically includes arrangements for a conference secretariat, either by creating a new temporary organization that reports directly to the General Assembly or by requesting one of the UN "specialized agencies" to provide appropriate logistical support.

---

[3]For more information about the debate surrounding the creation of a World Environment Organization see Biermann and Bauer (2005), Ivanova (2009).

A UN special conference is:

> A conference that is not part of the regular recurrent conference program of a biennium, but that is convened in a response to a specific resolution by the General Assembly of the Economic and Social Council. For whose substantive preparation and specific additional budgetary provisions are made and which all States are normally invited to attend [sic]. Such a conference usually extends over a period of a minimum of two to a maximum of four to six weeks and requires an intensive level of planning and servicing (Allen et al. 1982:1).

To date, the UN has hosted three environmental mega-conferences, with a fourth mega-conference scheduled for 2012. The "mega-conference" distinction arises from the conference's emphasis on the overall trajectory due to the complexity of the issue linkages between environment and development as well as the long lead times necessary for negotiations to come to fruition. Additionally, the mega-conferences' increased publicity status due to the heads of states in attendance distinguishes these meetings from more highly specialized single-issue negotiations. In addition to attracting heads of states, these meetings also convene thousands of other participants and observers. Pianta (2001:169) theorizes that the role and activities of the summits include framing the issue, rulemaking, policy guidelines, and enforcement. Willetts (1989:42) theorizes that mega-conferences attempt to change the global agenda in at least one of four ways: "adding a new issue; redefining the nature of an existing issue; giving more attention to an issue, possibly in order to resolve the issue; or creating a new institution, with its own formal agenda devoted to the issue."

Schechter (2005:6) points out that the rise of UN conference diplomacy occurred shortly after the increase in the number of developing countries in the UN General Assembly, who were more likely to favor this approach. Conference diplomacy to date tend to deal more with social and developmental issues. However, developing country support, in and of itself, is not a necessary condition for the convening of a conference, as funds separate from the UN budget must be raised. This normally entails the political and financial support of at least one of the industrialized countries. Mega-conferences are likely to be highly polarized along the North-South gap, including deep differences stemming from differing norms and values. Additionally, Taylor (1989:9) states that *ad hoc* conferences arise in part due to the lack of built-in procedures to identify and handle social issues at the UN, particularly within the ECOSOC. Consequently, *ad hoc* conferences deal with highly charged social and political issues that tend to suffer the political effects of the continuing North-South gap that characterizes much of UN politics.

Within each individual mega-conference planning cycle, the Preparatory Committees or PrepComs are responsible for deciding the agenda and negotiating text for the conference. PrepComs may also set rules for participation, asking for guidance from the General Assembly in the case of procedural deadlock. These arrangements have evolved gradually over time, as changes in communications and modes of travel have eliminated the necessity for conferences that meet for long lengths of time. Consequently, the "typical" conference time is 2 weeks, although several of the 1992 Earth Summit PrepComs lasted for longer periods.

Walker (2004) prefers to think of multilateral conferences as committees of states, represented by an individual or groups of individuals. Multilateral conferences vary in size ranging from small conferences with ten states to large conferences with 190 states and operate based on consensus. Additionally, the author believes that this setting emphasizes procedural rules and individual agency.

> [T]hat whereas power relations do not miraculously evaporate when delegates enter a conference room, in a multilateral conference factors intervene which are far less present in bilateral dealings between powerful and weak nations. There is power in effective argumentation, in skilful use of process and system, in the ability to assemble other delegations to support your position, etc (Walker 2004:247).

According to Walker, politically weaker states can manipulate the unusual rules and circumstances of the multilateral conference to concentrate power through superior use of knowledge and persuasion in order to change outcomes. Formally, states are the main participants at the PrepComs, but precedence dictates that NGOs registered with ECOSOC may also attend these meetings. That is not to say that all communications occur during the designated times for meetings or through the formal prescribed mechanisms for making oral presentations. In reality, the lead up to a conference involves communications not only among governments at the table, but also includes secretariat staff, NGOs, specialized agency staff, and the media.

It is perhaps more appropriate to think of a multilateral conference as a committee of professional diplomats who, in turn, represent a larger group of people based on geography (states), economic interests (business and industry, along with the NGOs that interceded on their behalf at the UN) or strongly-held beliefs (other NGOs, such as the ENGOs). There is no requirement for any individual to belong to only one diplomatic group. In fact, it is increasingly common that conference attendees will belong to more than one group, and therefore may have more than one competing identity at a conference.[4] Furthermore, analysts cannot assume that each attendee is a rational actor or that each person will always support the state at the expense of all others in which he/she has membership. Thus, in the mind of any given attendee, membership in a belief group may be more important than national identity. The multilateral conference gives states preferential treatment, in large part, because the state is the fundamental organizational structure of international society. Each individual is born into a specific state. One can change the state of residence and can even change national identity; it is not possible to live, however, outside of the state system. Membership in an organization, in contrast, is voluntary and changeable.

Thus, the multilateral conference functions as the basic building block of international treaty negotiation. That is, diplomats negotiate each individual treaty through a related series of conferences, or perhaps at a single conference, although environmental treaties have become so complex that multiple meetings are the norm. While these conference are not the only place where political activity occurs, they nevertheless occupy a privileged position because it is one of few instances when

[4]It has become commonplace and ordinary for states to include members of NGOs and businesses as part of the state delegation (Walker 2004:15).

diplomats create hard law. For example, the WSSD occurred as a series of five separate, though related conferences – the four preparatory conferences and the fifth headliner conference.

Procedurally, states participate in the main or official conference with NGOs acting as observers along with other important international bodies. In addition to the formal conference, the UN has made institutional arrangements for a parallel summit for NGOs, beginning with the UNCHE. NGOs utilize these parallel conferences to increase political pressure on the negotiations, often by increasing media attention, although other lobbying techniques such as drafting alternative treaty texts and highlighting new policy options also occur.

Despite the creation of UNEP as a result of the UNCHE, Stockholm did not directly give rise to sustainable development movement, the dominant contemporary international environmental philosophy, as has been subsequently claimed (Tolba 1998:2).[5] At the end of the Stockholm conference, it was not inevitable that sustainable development would become a dominant discourse, much less a regime, although the chain of events has not ruled out the possibility of a regime in the future. Unexpectedly, the historical trajectory of sustainable development has strong roots in the Stockholm NGO forum. What the Stockholm conference did do was to concentrate attention on the problems of environmental degradation and on potential solutions already underway to create a system of environmental governance under the aegis of national and intergovernmental authorities (UNEP 1982). This tradition of negotiating environmental treaties at *ad hoc* conferences has grown stronger in the years since Stockholm.[6] Unfortunately, this system also has led to the fragmentation of the environmental-issue arena, as each subset of environmental problems has its own conference and conference secretariat.

That is not to say that Stockholm stands alone as the only major environmental mega-conference. After a 20-year wait, it would have a worthy successor in the UNCED, more commonly known as the Earth Summit, which placed sustainable development at the heart of international environmental negotiations. This event continues to be the single largest conference on environmental affairs at the international level. The Earth Summit produced several important treaties and statements of principles including Agenda 21, the Framework Convention on Climate Change, and the Convention on Biological Diversity.

Agenda 21 contains policy prescriptives for implementing sustainable development by all levels of government, as well as by business and industry, and global civil society. This "blueprint for sustainable development" was crafted through a series of international negotiations as part of the preparatory meetings for the Earth

---

[5]While Stockholm created a negative linkage between environment and development, the more positive connotations of the term sustainable development did not enter the international arena until the 1980s. See Chap. 3 for a detailed treatment of the conceptualization of sustainable development.

[6]The tradition of an "automatic" review of a conference after 5 years ended in 2003, when the UN General Assembly stipulated that reviews of follow-up conferences would occur on an *ad hoc* basis.

Summit. Agenda 21 was organized by chapter and acceptance and implementation varies based upon the subject matter.

The Earth Summit served to install sustainable development as an integral, although separate, part of the international environmental agenda. In certain respects, it moved the concept of sustainable development closer to the Northern viewpoint of science-oriented environmental protection over the Southern concerns about development. Northern sensibilities about lifestyle issues – i.e., consumption – prevented the developing countries from advancing their quality of life agenda. An example of this battle occurred over Agenda 21's Chapter 4, which dealt with consumption. This chapter came to the Earth Summit entirely in brackets and negotiators eventually removed it from the document (Cohen and Murphy 2001; Lafferty and Meadowcroft 2000).[7] In this instance, the developed North triumphed over the developing South.

Compared to previous statements that touted developmental concerns over environmental protection, the Earth Summit was viewed with a great deal of cynicism and skepticism from its Southern originators. During the aftermath of the Earth Summit, the Southern bloc was able to propose new programs for developmental aid, but they could not force the developed countries to follow through on these commitments. For example, the United States and the United Kingdom immediately objected to proposals to expand UNEP's role into developmental affairs (Earth Negotiations Bulletin 1991). Technology transfer also yielded two arenas of conflict – intellectual property rights and concessional versus commercial terms for technology transfer.

The Earth Summit finalized the split in the environmental agenda between technical environmental items and normative issues. While the climate change and biodiversity treaties unquestionably bolstered the meeting's importance, these two treaties also signified a move to keep technically oriented treaties away from the more value laden ad hoc conferences. Of the five broad issue areas – climate change, biodiversity, desertification, deforestation, and sustainable development, only the last carried over to the WSSD in Johannesburg. That is not to say that the other areas disappeared from the international agenda. Instead, they continued to grow in importance as evidenced by their own evolving conference tracks. The Kyoto Protocol and the Marrakech Accords combined to give considerable importance to the climate change regime, effectively creating a regime despite the lack of participation by the United States.[8] The biodiversity treaty was followed by the Cartagena Protocol on Biosafety that stipulates the conditions and terms of use, including legal liability and redress, for genetically modified organisms that are intentionally introduced into the environment. The Cartagena Protocol also established the Biosafety Clearinghouse to provide technical assistance with program implementation as well as to serve as an information exchange on genetically modified organisms.

---

[7]Bracketed text indicates countries inability to agree on the formal wording of a treaty.

[8]For more information on the negotiating history of the Kyoto Protocol, see Oberthur and Ott (1999) and Yamin and Depledge (2004).

In keeping with the tradition of holding a major environmental conference every 10 years, the UN convened the WSSD in 2002.[9] Unlike the Earth Summit, the Johannesburg conference could hardly claim the same heady importance and the UN was hard pressed to avoid declaring the event an outright failure. In part, the WSSD suffered from bad timing as it occurred in the aftermath of the September 11th attacks in the United States. However, the conference also suffered from poor organizational skills, a lack of preparatory work, and an absence of adequate finances to hold the initial PrepComs. Given this lackluster beginning, it is perhaps not surprising that the WSSD generally failed in its objective to further states' implementation of the sustainable development agenda. Instead, the Johannesburg conference highlighted the erosion of support for Agenda 21, the policy prescriptives negotiated at the Rio Earth Summit.

While the UN General Assembly undoubtedly wanted Johannesburg to be an action-oriented conference, the geo-political status at the time gave little reason to believe that any meaningful results would be achieved. The North-South financial gap had grown wider, environmental pollution continued at unprecedented rates, and foreign aid decreased, both in real terms and as a percentage of GDP. The United States, in particular, had undergone a major shift in foreign policy with the change of its presidency from President Clinton to President Bush. As such, it was unlikely to support any agreement that might come out of the WSSD process.[10] Further, the world's one remaining superpower was fighting a war with Afghanistan after suffering from a devastating attack at home in its commercial center, New York, and its political capital, Washington D.C. The September 11th attacks on the World Trade Center and the Pentagon served to further lower American interest in attempting to change the socio-economic system it was currently defending through military action.

The WSSD in Johannesburg produced two significant documents —the Declaration on Sustainable Development and the Plan of Implementation. Overall, the Summit bolstered sustainable development by repeating reassuring words of governmental support. However, governments attending the conference were unwilling to make meaningful long-term commitments to this international ideal. One major success for the environmental movement occurred with the failure to designate the World Trade Organization (WTO) as the appropriate forum for reconciling conflicts between environmental treaties and trade agreements. In a more cynical vein, the fact that this item was on the agenda could, in and of itself, be viewed as a major setback for pro-environment conferees.

The inclusion of Type 2 partnerships as a major conference outcome demonstrated a failure by states to make significant headway on Agenda 21 in the past 10 years. Type 2 partnerships explicitly attempted to pull non-state actors into the

---

[9]The UN established this precedence with the first follow-up "conference" – a UNEP Governing Council of a Special Session that occurred in Nairobi in 1982.

[10]One of President Bush's first major foreign policy decisions was to withdraw the United States from the Kyoto Protocol. For more information about this decision, see Cohen and Egelston (2003).

sustainable development discourse by proclaiming these groups a partner alongside governments. The partnerships heralded a shift of emphasis from state action to voluntary action by elevating the importance of NGOs and other kinds of non-state actors within sustainable development. Consequently, these non-state groups became an explicit target of the negotiations. To a certain extent, the partnerships were designed to hide the fact that the Johannesburg conference was largely unsuccessful. They were born of a necessity for the UN to deliver some kind of financial assistance to the developing South, but it generally is not clear whether or not the funds these non-state groups promised are "new and additional."

Norris (2005) performed a preliminary analysis on the Type 2 partnerships negotiated as part of the WSSD process. Her report shows a surprising lack of participation from certain actors, most notably the low presence of businesses and the absence of China and India. Partially as a result of this, the partnerships themselves came under increased criticism. By the end of the meeting, corporations, in particular, had shied away from publicizing new commitments.

The most notable difference between Rio and Johannesburg involved the greater embeddedness of social justice and economic development at the expense of environmental protection. For example, the Johannesburg Declaration mentions "the need for human dignity for all", "the indignity and indecency occasioned by poverty", and "the need to produce a practical and visible plan to bring about poverty eradication and human development", before mentioning natural resources, biodiversity loss, desertification, and climate change. This language signals yet another shift in the meaning of sustainable development away from the North's preference for environmental protection and toward the South's preference for industrial development.

As a point of comparison to the earlier conferences, the "quality of life" embedded within environmental affairs as a result of the Stockholm Conference slowly gave way to a more technically oriented set of goals – access to potable water, improvement in sanitation systems, and housing that the developing countries frequently lack. While these easily quantifiable goals are laudable, the likelihood of developed countries increasing funding to provide these necessities outside of their borders is perhaps unrealistic. It is politically unpopular to increase development aid given the graphic inequalities of wealth distribution inside the United States as pointed out by the "Occupy" protests that started in 2011 and continue in 2012.

This combination of factors – a lack of political leadership and the failure to secure new financial commitments to sustainable development in the wake of the 2001 terrorist attacks led to a lack of consensus about the future of Agenda 21 at the end of the Johannesburg meeting. WSSD concluded with the hopes of an increased state commitment to sustainable development temporarily dashed, but with some slight increase in commitment to sustainable development from the business and industry community.

Despite multiple UN conferences, most prominently UNCED and WSSD, that sought to define and implement sustainable development, international lawyers note that sustainable development has yet to become a customary norm, and as such is not a part of binding international law (Boyle and Freestone 1999; Handl 1998).

Nevertheless, a significant coalition of states, global civil society, and businesses increasingly act in accordance with the norm of sustainable development even though they have selectively implemented its main documents. That is, there is an important distinction between sustainable development as a set of normative concerns and sustainable development equated with the implementation of Agenda 21.

2.3 NGOs at the UN Conferences

In layman's terms, an NGO is an organization that is neither a state nor an intergovernmental organization (Feld and Jordan 1983). Because states dominate the UN, business groups must also organize themselves into NGOs for the purposes of observing UN meetings. Prominent business organizations observing UN affairs include the ICC and the World Business Council on Sustainable Development.

As a point of historical fact, a number of NGOs predate the creation of the UN. While identification of the first NGO to involve itself in international politics is a matter of debate, all factions agree that civil society groups that are the precursors to today's NGOs first began to appear between 1840 and 1870 and included the International Red Cross and the Young Men's Christian Association (Charnowitz 1997; Chiang 1981; League of Nations 1938; Stošić 1964; White 1951). The International Red Cross was particularly important as it won a Nobel Peace Prize for its humanitarian work during World War I. For the most part, though, these organizations slowly grew in number and in political power, particularly by interacting with the League of Nations. This relationship with the League was due in part to the hesitancy of states to support the work of the League. The more politically astute NGOs were able to utilize this power void to promote their own groups. However, World War II interrupted the growth of these groups. Despite the presence of a significant NGO contingency at the San Francisco conference that established the UN (the United States invited 42 NGOs), only four groups were accredited to the UN in 1946 (Willetts 1996a:33).[11] The growth of NGOs occurred slowly through the early years of the UN, with approximately 298 NGOs registering to attend the "official" NGO parallel conference at UNCHE in 1972.[12]

The UN enforces strict rules in determining which organizations may attend meetings at the world body. In the language of the UN, an NGO is "a not-for-profit, voluntary citizens' group, which is organized on a local, national, or international

[11]The UN Yearbook (1947) indicates that the first four NGOs accredited to the UN are the American Federation of Labor, the ICC, the International Co-Operative Alliance and the World Federation of Trade Unions.

[12]Multiple other NGO conferences occurred during the same time as UNCHE in Stockholm. For a complete description of these other events, see Chap. 4.

level to address issues in support of the public good" (UNDPI 2004).[13] Four prohibitions on NGO activities exist, by definition. NGOs may not (1) be a for-profit corporation, (2) engage in widespread, politically-motivated violence, (3) seek to overthrow legitimate state governments, or (4) work against the international organization that grants them status (Willetts 1996b:3).

For the most part academics, as well as some NGOs themselves, have rejected this definition. The claim is that the organizations typically grouped in this category are too diverse to allow for careful study and theorizing. In addition, various NGOs are ideologically opposed to each other (Willetts 1996a:11). Progressive NGOs that have seen themselves as fighting against the close relationship businesses supposedly enjoy with government, have deeply resented joining their ranks. Likewise, scholars simplified the analysis by assuming that ENGOs will always oppose BINGOs; thus, non-business NGOs are "true" NGOs and BINGOs are often treated as "evil" pretenders. Accordingly, many academic works restrict NGOs to nonbusiness, nonprofit groups in order to allow for theoretical clarity, as there are significant differences between these two subgroups (Chatterjee and Finger 1994; Lindborg 1992).

On the other side of the debate, restricting NGOs by subgroup artificially creates a distinction in the scholarly literature that is not representative of current practice. For example, this approach dismisses BINGOs as irrelevant, when in fact the ICC was one of the original NGOs granted consultative status with ECOSOC in 1946 (UN 1947:554). It also creates a linguistic distinction between practice and theory. The scholarly argument that demotes the ICC from the realm of NGOs is ignominious considering the rich tradition and history of this group.[14] Careful study of BINGOs remains a recent development within international relations (Lisowski 2004).

For the most part, this text will reluctantly follow the scholarly convention of restricting NGOs to those organizations that support the sustainable development discourse. This is more a matter of convenience than anything else. Information about business groups over the course of this case study is generally absent. Further, many scholarly theories utilize this convention, and agreeing with this definition simplifies the theoretical analysis later in this book.

To complicate matters further, the definition of a NGO has changed over time. Originally, the UN required NGOs to be "international." Internationalism versus single nationality was an important determinant in conferring legitimacy among NGOs. The distinction gradually fell out of use during the 1970s, although Feraru

[13]The classic definition of public good states that it is a good or service that is both non-rivaled and non-excludable. That is, everyone can consume the good without diminishing the usefulness of the good and that individuals are not easily excluded from consuming the good. Clean air is frequently cited as an example of an environmental public good. There is a substantial literature on public goods, frequently referred to as common pool resources. See for example, Barkin and Shambaugh (1999), McCay and Acheson (1987), McKean (2000), Olson (1965), Singh (1994).

[14]The ICC is one of a handful of NGOs that was invited by states to sign an international economic agreement, occurring once in 1927 and again in 1928 (White 1951:21).

(1974:33) used it as a criterion in her research into the influence of NGOs at Stockholm.[15] Prior to this time, a litmus test of at least six nationalities had been proposed and sometimes applied as a necessary characteristic for UN recognition. The most recent guidelines for NGOs, ECOSOC Resolution 1996/31, have dropped the "international" requirement, allowing both national and local NGOs to participate in the multilateral conferences. However, scholars sometimes distinguish between "local" and "international" NGOs by requiring that NGOs have a "global" impact as a litmus test for inclusion (Arts 1998:50).

ECOSOC, which oversees the work of the UN on economic and social issues, including human rights and the environment, was proposed along with the other constituent elements of the UN system at Dumbarton Oaks in 1944. However, ECOSOC was not part of the proposed structure for the UN. Additionally, at the time of its creation, NGOs had no formal role at the UN. Further, there is a clear intent dating back to the San Francisco Conference to deny NGOs access to the new world body. In fact, it was only at the last minute, and with well-placed domestic pressure, particularly from the United States, from groups that had status under the old League of Nations system, that NGOs were able to secure any representation within the UN. The United States government changed its position in response to NGO pressure at the 1945 San Francisco conference that provided the UN charter, clearing the way for the arrangement that exists today. The UN Charter in Article 71 stated:

> The Economic and Social Council may make suitable arrangements for consultation with non-governmental organizations which are concerned with matters within its competence. Such arrangements may be made with international organizations and, where appropriate, with national organizations after consultation with the Member of the United Nations concerned (UN 1945).

Even so, NGO representation, in the form of providing apolitical technical advice, was limited to the lesser bodies of the UN such as ECOSOC, instead of the General Assembly or the Security Council.

Interim arrangements for ECOSOC-NGO consultations were set forth in ECOSOC resolution 2/3 in 1946. ECOSOC spent the next 5 years working to finalize this requirement, culminating in the issuance of ECOSOC Resolution 288 B(X) in 1950. NGOs were divided into three categories based on their competence on the issues expected to be brought before ECOSOC: Category A included organizations with "a basic interest in most of the activities of the Council"; Category B covered NGOs with "a special competence"; while Category C organizations focused on "the development of public opinion and with the dissemination of information" (UN 1947:552).[16] Other specialized agencies were

---

[15]Resolution 1296 (XLIV) in 1968 allowed national NGOs to apply for consultative status although it took several years before the international/national distinction disappeared (UN 1968).

[16]Category C officially became the Registry in 1950.

free to determine their own relationship with NGOs, although access to information and accreditation practices tended to conform to the ECOSOC regulations in place at the time.

ECOSOC revised its relationships to the NGO community in May 1968 after *The New York Times* published information indicating that the United States Central Intelligence Agency was covertly funding certain NGOs.[17] The 1968–1969 review of NGOs was a heated debate that reflected a number of changes that had occurred in the UN since its inception in 1946, most notably the increase in the number of developing countries and Eastern European states and the expansion of UN bureaucracy that inevitably accompanied the growth in the number of member states. The review also brought to light the extent to which NGOs participated in the daily affairs of the UN, particularly in areas of low politics – human rights, development, and the environment.

The change in composition within the states of the UN, meant, among other things, that NGOs could no longer take for granted that the member states of the UN were in favor of democratic principles and traditions, including the right to criticize freely government officials' decisions and actions. Consequently, member states forced NGOs to retreat from their frequent and vocal criticism, especially in humanitarian affairs, of member states. Additionally, NGOs documented their activities to the developing countries, including the provision of knowledge and funding, either from private donors or from Western governments (Chiang 1981:198). In other words, NGOs found themselves having to promote their "technical expertise" while downplaying their "political" nature.

In practice, ECOSOC Resolution 1296 changed the pre-existing resolution governing NGO relations by changing the names of Category A and B to Category I and II, while the Register was renamed the Roster. The UN allowed all NGOs to attend ECOSOC meetings, as well as to circulate written statements as official UN documents. Additionally, Category I NGOs could introduce items on the ECOSOC agenda, while both Category I and II NGOs could petition to speak to the Council, although the procedures and length of speeches vary between the two categories (Category I circulated slightly longer documents and could speak for longer amounts of time). More importantly, the Categories created a perception of hierarchy among governments and NGOs. NGOs were also required to update their information every 4 years and a new procedure for withdrawing consultative status was instituted, although the practice had, in fact, happened in the past, albeit infrequently. In the extremely rare case where the UN withdrew consultative status, the NGO failed to file the appropriate paperwork.

The 1968 changes stayed intact until 1996. In that year, the UN General Assembly fulfilled a mandate in Agenda 21 negotiated at the UNCED in Rio

[17] Certain government officials saw funding of NGOs during the cold war as a flagrant violation of international protocol, although many NGOs receive government funding, albeit not for spying, today. For more information, including bibliographic references to the original New York Times articles, see Chiang (1981:77).

de Janeiro. Chapter 23 of Agenda 21 stated that the UN would reexamine its relationship with NGOs. This review resulted in two major changes. First, the UN ended the "international" NGO requirement, formalizing the gradual opening of the UN to first national, than regional and local NGOs. Second, the UN expanded its roster to a large number of small organizations interested only in environmental affairs such as the Centre for Our Common Future. Initially, the Secretariat created a separate category for NGOs that wished to attend the Commission on Sustainable Development, also created at the 1992 Rio Conference. However, the UN allowed organizations on the "CSD roster" to apply for full status and the CSD roster became simply the roster. Lastly, the ECOSOC roster changed the names of its categories. Category I became General Status, Category II became Special Status. While the change in terminology did not alter the classification or privileges associated with the appropriate categories, the name change did result in considerable confusion. A table highlighting the changes in terminology is below.

1950	1968	1996
Category A	Category I	General status
Category B	Category II	Special status

The confusion stemming from the 1996 rule changes resulted, in part, from the significant growth of NGOs that had registered under the CSD roster as a result of Rio's Global Forum. Willetts (1996a:38) provides the most widely cited details of the NGO sector. He analyzed the growth of NGOs based upon the information provided in the UN Yearbook from 1945 to 2002. The graphical format of this data is presented in Fig. 2.1. Willetts dataset shows a dramatic increase in the growth of NGOs from 1945 onwards. However, this data set does not represent the total NGO universe for three reasons. First, Willetts did not investigate why the changes in NGOs occurred, except in the instances of a UN ECOSOC rule change. For example, 1979 shows an overall decline of 151 NGOs from 1978. Willetts fails to distinguish whether or not the NGOs "died" or whether the NGOs were removed from the UN roster through a change in procedural reporting or through the forces of bureaucratic inertia. Second, Willetts dataset consists only of NGOs willing to register with the UN. NGO registration requires NGOs to legally exist for at least 2 years prior to registration, which skews the dataset. For example, an NGO may form in 1999, but may not apply for ECOSOC registry until 2002 at the earliest. Lastly, none of the UN environmental conferences limited conference participation to the NGOs formally registered with ECOSOC. Each conference allowed NGOs lacking ECOSOC registry to request accreditation directly from the conference secretariat. These limitations do not preclude the usefulness of the dataset, per se. A complete listing of NGOs has never been accomplished, nor, given the rate of proliferation of NGOs, is it likely to occur in the future. While the ECOSOC date of formation may have at a minimum, a 2-year time lag, the general shape of the trend, the specific concern for this research, does not change.

The scholarly differences in opinion as to the reason for the formation and spread of NGOs, occur in part, because scholars examined the formation of NGOs in

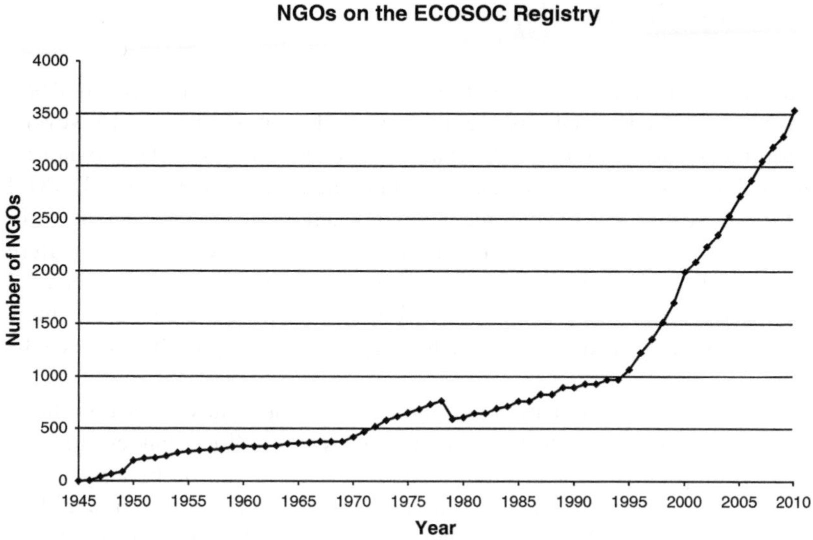

Fig. 2.1 NGOs Appearing on the ECOSOC Registry

different geographical locations at different times. Consequently, both approaches have explanatory power, albeit in slightly different situations. Originally, western grassroots organizations formed in the 1960s and 1970s because of strong local interest. However, these new groups had to find a niche to fill in order to continue justifying their existence and fundraising efforts. As this book shows in later chapters, environmental organizations found one niche in providing consultative services to various parts of the UN system, including the newly created UNEP after Stockholm in 1972. Initial successes in this political opportunity in turn justified further local activity and fundraising in a symbiotic relationship as evidenced by the establishment of the Environmental Liaison Centre International (ELCI) in Nairobi in 1974.

Consequently, as the UN system expanded further into international environmental affairs, the NGOs grew along with the UN-led expansion into "social" issues such as human rights, environment, and development. The Northern governments, along with the UN system and private funding sources, gave substantial financial assistance and advice to both established groups and interested individuals in the South in order to organize pro-environmental groups in the Third World (Wapner 1996), along with other humanitarian organizations (Reimann 2006). Additionally, both these new NGOs and existing NGOs proved willing and able to provide additional services to the UN as environmental affairs grew in complexity. At the same time states began to evince an apparent inability to muster both the political will and institutional support to implement successfully all of the changes in behavior suggested by new international norms under discussion at the UN. As a result, NGOs willing to provide this support were welcomed into the UN system.

2.4 Sustainable Development

Sustainable development slowly emerged as the overarching environmental philosophy in international negotiations over the course of the past 25 years. However, what this term means and why it should matter to the citizen on the street is not readily apparent, despite the lip service paid to sustainable development at the UN. Multiple definitions of sustainable development abound, and more enter the debate with each passing publication. These definitional differences arise due to cultural differences as well as in differences in economic needs or rights, both real and perceived. Even the attempts to count the number of definitions for sustainable development has proven meaningless since Murcott (1997) reported finding 57 distinct uses for the term.

Sustainable development arose out of the need to integrate concerns of the underdeveloped countries into international environmental politics, namely to reconcile environmental protection with future economic growth. At the first environmental conference, it was assumed that environmental protection would significantly retard economic growth and this created widespread opposition in both the North and South. Sustainable development originally sought to overcome this problem by focusing on ways in which economic growth could occur without environmental damage. It gained political traction slowly during the 15 years after the Stockholm conference in 1972, until one of three commissions focusing on various aspects of North-South relations in the early 1980s promoted the concept.[18] The WCED provided the most famous definition of sustainable development in its report, *Our Common Future*, published in 1987. Sustainable development is "development that meets the needs and aspirations of the present without compromising the ability to meet those of the future" (WCED 1987:40). As such, sustainable development intertwines economic growth, social justice, and environmental protection. It seeks to ameliorate poverty, yet simultaneously to conserve the natural resources that undoubtedly will be consumed in the process of improving the quality of life in the developing countries, particularly in the least developed countries. Further, sustainable development seeks to expand economic growth while simultaneously celebrating diversity and expanding public participation (Cordonier Segger and Khalfan 2004:3). In doing so, sustainable development calls for equal access for all; not only to natural resources, but also equal access to transparent political processes. Sustainable development deals in complexity as it also ensconces multiple political and normative positions including common, but differentiated responsibilities, development assistance, technology transfer, distribution of wealth – both between nations and within states, and intergenerational equality.

[18]The other two commissions are the Brandt Commission, more formally known as the Independent Commission on International Development and the Palme Commission, the Independent Commission on Disarmament and Security Issues. The Brandt Commission issued two reports *North-South: A Program for Survival* and *Common Crisis: North-South Cooperation for World Recovery*, while the Palme Commission issued one report – *Common Security*.

The WCED political mandate included proposing a "long-term strategy for sustainable development" that would lead to increased cooperation between nations and people with respect to common objectives about meeting the needs of the poor while "protecting and enhancing the environment" (WCED 1987:ix). Langhelle (1999) argues that *Our Common Future* is an internally consistent document that was much more radical than critics recognized in that the report sought to promote long-term development, even at the expense of certain environmental protection regulations. In the words of Sachs (1993:9), *Our Common Future* called for the "conservation of development" not the wholesale conservation of the environment. In other words, certain environmental degradation was acceptable if it could help meet the legitimate needs of the world's poor.

Lélé (1991) traces the semantic meanings proposed by different sustainable development advocates. In doing so, she also provides an overview of the conceptual history of the term. The term sustainability emerged from forestry and fisheries management and referred to the resource renewal rate. Theorists and practitioners frequently conflate development with gross national product (GNP) growth, although in its most basic form development refers to a deliberate process of growth. Combining the two terms would suggest that sustainable development refers to economic growth that does not cause resource depletion over time. In doing so, sustainable development refers to a specific normative orientation. Notably, sustainable development does not point to a specific set of prescriptive behaviors for various governmental agencies and private actors to adhere to, although Agenda 21, drafted by national governments at the 1992 Earth Summit, attempted to do so.

In the years since the Earth Summit, sustainable development as a normative concept has been partially decoupled from the implementation of Agenda 21. Participants at the Rio conference stated a great hope that the implementation of Agenda 21 would give impetus to the widespread social and economic changes envisioned as part of sustainable development. However, upon further examination, the policies and principles contained within Agenda 21 proved intractable to implement at the national level, and the policy document lost its prominent position not only in rhetorical speeches, but also in regulatory circles. That is not to say that sustainable development declined during the same time. Quite contrarily, support for sustainable development has slowly increased over time, despite the fact that its main policy document failed in its purpose to provide clear policy guidance to states.

However, the radical implications of the "conservation of development" have not entered the mainstream debate as environmentalists, along with other neo-liberal forces moved the use of the term towards a synonym for limiting climate change and biodiversity loss. The importance of this connection stems not only from the linkage between biodiversity and climate change through the role of plants and trees via the carbon cycle, but also from a deep-seated suspicion that humankind cannot be certain whether or not the plants and animals being systemically destroyed could prove highly useful in the future.

This issue linkage proved to be a setback in the long term as climate change and biodiversity continue to be highly controversial within the United States and therefore contributes to political resistance to sustainable development. Consequently, sustainable development remains both a contested term and a contested norm. There is, of course, a significant political advantage arising from competing meanings. The ambiguity masks deeply divisive policies, allowing the appearance of agreement and cooperation, when in fact radically different discourses exist (Dryzek 2005).

It would be misleading to paint an overly positive acceptance of *Our Common Future* as the report also unleashed a hailstorm of criticism, particularly from ecologists, who rejected the principal of future unlimited economic growth (Clow 1990:7; Lohmann 1990; Rees 1990:18). Likewise, certain organizations in the developing countries feared that sustainable development was, at best, another new rhetoric for detracting attention from their problems, or, at worst, a deliberate attempt to subvert national control by increasing funding for the failed development schemes in a futile attempt to "fix" the problem (Wagle 1993:317).

Yet outright dismissal of sustainable development as a polite, but ultimately futile, gesture of political idealism fails to recognize the substantial contributions that the concept has made to soft law. It is instead perhaps more prudent to recognize sustainable development as a set of discursive practices with significant political support, particularly in Western and Northern Europe (Norway and the Netherlands) and parts of the developing world (Lafferty and Meadowcroft 2000). Its powerful political allies also extend into the international financial development banks including the World Bank and the Asian Development Bank, along with other political organizations including the Organization for Economic Cooperation and Development and the North Atlantic Treaty Organization (Lélé 1995:227). Most importantly, the UN remains sustainable development's most dedicated advocate. That is not to say that sustainable development enjoys unanimous support, as significant impediments to the broader adoption of both the norm and the regime exist, most problematically the United States' steadfast refusal to engage meaningfully in discussions on this topic.

Outside of the state system, environmental organizations were able to adopt, and afterwards to promote this philosophy. They viewed it as an outgrowth of the earlier environmental paradigms of conservation and soft energy prominent in the 1970s (Buttel et al. 1990:61). NGOs' political support for this paradigm manifests itself in a number of ways ranging from support to international meetings discussing sustainable development to the formation of partnerships with private industries looking to adopt sustainability goals.

At the same time, BINGOs utilize sustainable development as a tool for differentiating one business from the other, an important intangible benefit for both stakeholders and consumers. Corporate sustainability reporting continues to grow rapidly and it is now impossible to tell who the "greenest" company is due to these competing claims. Companies oft mentioned in the running, however, include 3M, Johnson and Johnson, IBM, and Google. In contrast, ExxonMobil utterly repudiates this stance and claims for itself to be acting not only in its shareholders best interests, but also by contributing to Adam Smith's "invisible hand" that distributes social benefits within the marketplace. Consequently, whether a company or organization is for or against sustainable development, the designation has usefulness in distinguishing actors at the global level and is unlikely to disappear.

The question, then emerges, what political form does sustainable development take? Several possibilities can be immediately eliminated. Sustainable development is not hard law as it is not supported by a binding treaty with enforcement provisions. Likewise, sustainable development is not a regime. Krasner (1982:186) defined a regime as "explicit principles, norms, rules, and decision-making procedures". This definition is sufficiently broad to leave open the possibility that sustainable development could fall into this category, but Strange (1982) and others have handily criticized this definition as too vague to be useful. Additionally, later regime theorists such as Young (1997) have moved the definition of regime closer to that of an institution. Multiple intergovernmental organizations could serve as an institution for the sustainable development regime, but in reality there is no intergovernmental institution focusing on sustainable development mandates or specific actions to be taken, although several intergovernmental organizations might make suggestions as to which actions might be prudent.

More problematically, actors that do not adhere to the principles of sustainable development are not subject to disciplinary action from the rest of the international community, although actors and businesses that do attempt to follow the tenets of sustainable development are publicly praised. Lack of enforcement is one of the key signals that a subject area is not a regime. It is possible that sustainable development could become a regime in the future. Thirty years is not an unreasonable amount of time for establishing a regime at the international level, particularly when the issue area is complex, the changes required are substantial, and the resistance to the movement is significant. In short, sustainable development is very different from the early days of environmental protection that sought to clean air and water by adding equipment to clean substances prior to entering the environment.

Rather than define sustainable development as a regime, sustainable development could be described as a discourse. Litfin (1994) defined discourse as "sets of linguistic practices and rhetorical strategies embedded in a network of social relations." Scholars such as Dryzek (2005) and Redclift (1987) have documented the differing discourses that utilize the term sustainable development. While sustainable development is a definition whose meaning (along with normative content and thereby actions necessary) is contested, there is nevertheless some minimum requirement for acting in a manner consistent with sustainable development. Keck and Sikkink (1998) state, "Norms and practices are mutually constitutive – norms have power in, and because of, what people do." Norms are powerful because they cause the believer to change behavior. Meanings and actions that move beyond this minimum normative threshold are not of concern to supporters of sustainable development. Those that claim to be sustainable, but fail to meet this standard are decried as "greenwash." While it is something of an oversimplification to consider sustainable development as one unified discourse, the power of the norm is nevertheless apparent in the change of behavior that links norms to actions. In this respect, then it is appropriate to speak of sustainable development as constituting a single discourse, adoption of the norms of sustainable development causes a change in behavior that can be observed.

Likewise, numerous countries took on the implementation of sustainable development as specified in the Earth Summit documents. National workshops and local Agenda 21 Committees were launched throughout the world. That is not to say that states universally adopted sustainable development. On the basis of a cross-national comparative project of the uptake of sustainable development in nine developed countries and the European Union, Lafferty and Meadowcroft (2000) report that states vary greatly in their support of the concept and range from the "enthusiastic" support of Netherlands, Norway, and Sweden to the "disinterested" United States.

While it is theoretically appealing to proclaim that all NGOs support sustainable development, and businesses and industry organizations, as well as businesses themselves, oppose sustainable development, this approach oversimplifies a more complex reality. Murphy and Bendell (1999:5) recognized that both NGOs and businesses interests on sustainable development "remains exceedingly diverse, largely disorganized and quite often divided." This situation remains true today, although support for sustainable development has generally grown among these groups. They then go on to identify support from ENGOs, development NGOs, and Southern NGOs.

Business opposition to sustainable development has lessened over time as corporations have found ways to easily comply with elements of the sustainable development discourse. Compliance with national environmental, health, and safety regulations are increasingly reported as support for sustainable development. Further, certain progressive corporations determined that supporting sustainability is good for business. A company may produce an environmentally-friendly product. Additionally, promoting sustainable development now could lead to larger markets for their products later.

Additionally, businesses organized business groups in support of sustainable development, most notably the World Business Council on Sustainable Development (WBCSD). These organizations share case studies about best practices and help to raise awareness and support for the sustainable development discourse.

That is not to say that business uniformly support sustainable development, they do not. Further, businesses that do support portions of the sustainable development discourse may not support all of the elements of the sustainable development discourse. Businesses generally have no interest in elements of sustainability such as limits to consumption that may negatively impact their current or future business prospects.

Additionally, these groups may not support an identical sustainable development discourse. One of the hallmarks of a discourse is that meaning continues to evolve. Consequently, one of the impacts of the diversity of groups supporting sustainable development is also a diversity of underlying norms and supporting actions. Regardless of the strength of NGOs' support for the sustainable development discourse, these actors can no longer be ignored in attempts to preserve our beautiful blue earth.

References

Abbott, K., & Snidal, D. (2000). Hard and soft law in international governance. *International Organization, 54*(3), 421–456.

Allen, M. E., et al. (1982). *Report of the Joint Inspection Unit on Secretariat Organization and procedures for the preparations of the United Nations Special Conferences*. (JIU/REP/82/2).

Arts, B. (1998). *The political influence of global NGOs: Case studies on the climate and biodiversity conventions*. Utrecht: International Books.

Barkin, J. S., & Shambaugh, G. E. (1999). *Anarchy and the environment: The international relations of common pool resources*. Albany: State University of New York Press.

Biermann, F., & Bauer, S. (Eds.). (2005). *A world environment organization: Solution or threat for effective international environmental governance?* Aldershot: Ashgate.

Boyle, A., & Freestone, D. (1999). *International law and sustainable development: Past achievements and future challenges*. Oxford: Oxford University Press.

Buttel, F. H., Hawkins, A. P., & Power, A. G. (1990). From limits to growth to global change. *Global Environmental Change, 1*(1), 57–66.

Caldwell, L. K. (1984). *International environmental policy. Emergence and dimensions*. Durham: Duke University Press.

Caroll, J. E. (1988). *International environmental diplomacy*. Cambridge: Cambridge University Press.

Charnowitz, S. (1997). Two centuries of participation: NGOs and international governance. *Michigan Journal of International Law, 18*(2), 183–286.

Chatterjee, P., & Finger, M. (1994). *The earth brokers. Power, politics and world development*. London: Routledge.

Chiang, P. (1981). *Non-governmental organizations at the United Nations*. New York: Praeger.

Clow, M. (1990). Sustainable development won't be enough. *Policy Options, 11*(9), 6–8.

Cohen, M. J., & Egelston, A. (2003). The bush administration and climate change: Prospects for an effective policy response. *Journal of Environmental Policy and Planning, 5*(4), 315–331.

Cohen, M. J., & Murphy, J. (2001). *Exploring sustainable consumption: Environmental policy and the social sciences*. Oxford: Elsevier Science.

Cordonier Segger, M. C., & Khalfan, A. (2004). *Sustainable development law principles, practices and prospects*. Oxford: Oxford University Press.

Dryzek, J. S. (2005). *The politics of the earth: Environmental discourses* (2nd ed.). Oxford: Oxford University Press.

Earth Negotiations Bulletin. (1991). *Institutions* (Working Group III, Vol. 1, No. 20). March 27, 1992.

Feld, W. J., & Jordan, R. S. (1983). *International organizations. A comparative approach*. New York: Praeger.

Feraru, A. T. (1974). Transnational political interests and the global environment. *International Organization, 28*(1), 31–60.

Handl, G. (1998). The legal mandate of multilateral development banks as agents for change towards sustainable development. *American Journal of International Law, 92*(4), 642–665.

Hoggart, R. (1996). UNESCO and NGOs: A memoir. In P. Willetts (Ed.), *The conscience of the world* (pp. 98–115). Washington, DC: The Brookings Institution.

Hurrell, A., & Kingsbury, B. (1992). *The international politics of the environment. Actors, interests, and institutions*. Oxford: Clarendon.

Ivanova, M. (2009). UNEP as anchor organization for the global environment. In F. Biermann, B. Siebenhüner, & A. Schreyögg (Eds.), *International organizations in global environmental governance* (pp. 151–173). New York: Routledge.

Keck, M. E., & Sikkink, K. (1998). *Activists beyond borders: Advocacy networks in international politics*. Ithaca: Cornell University Press.

Krasner, S. D. (1982). Structural causes and regime consequences: Regimes as intervening variables. *International Organization, 36*(2), 185–205.

Lafferty, W. M., & Meadowcroft, J. (2000). *Implementing sustainable development strategies and initiatives in high consumption societies*. Oxford: Oxford University Press.

Langhelle, O. (1999). Nature, market and ignorance: Can development be managed? In W. Lafferty & O. Langhelle (Eds.), *Towards sustainable development: On the goals of development and the conditions of sustainability* (pp. 111–136). New York: St. Martin's Press.

League of Nations. (1938). *Handbook of international organizations (associations, bureaux, committees, etc)*. Geneva: League of Nations.

Lélé, S. M. (1991). Sustainable development: A critical review. *World Development, 19*(6), 607–621.

Lélé, S. M. (1995). Sustainable development: A critical review. In K. Conca, M. Alberty, & G. K. Dabelko (Eds.), *Green planet blues: Environmental politics from Stockholm to Rio* (pp. 227–238). Boulder: Westview Press.

Lindborg, N. (1992). Nongovernmental organizations: Their past, present and future role in international environmental negotiations. In L. E. Susskind, E. J. Dolin, & J. W. Breslin (Eds.), *International environmental treaty making* (pp. 1–25). Cambridge: Program on Negotiation Books.

Lisowski, M. (2004). *A three-level game analysis of the international climate change negotiations 2000/2001*. Dissertation. University of Cambridge.

Litfin, K. (1994). *Ozone discourses*. New York: Columbia University Press.

Lohmann, L. (1990). Whose common future? *The Ecologist, 20*(3), 82–84.

McCay, B. J., & Acheson, J. M. (1987). *The question of the commons: The culture and ecology of communal resources*. Tucson: University of Arizona Press.

McKean, M. A. (2000). Common property: What is it, what is it good for, what makes it work? In C. C. Gibson, M. A. McKean, & E. Ostrom (Eds.), *People and forests: Communities, institutions and governance* (pp. 27–52). Cambridge: MIT Press.

Murcott, S. (1997). *Sustainable development: A meta-review of definitions, principles, criteria indicators, conceptual frameworks and information systems.* Annual Conference of the American Association for the Advancement of Science. IIASA Symposium on "Sustainability Indicators." Seattle, WA. February 13–18, 1997.

Murphy, D. F., & Bendell, J. (1999). *Partners in Time? Business, NGOs and Sustainable Development* (Discussion Paper 109). Geneva: UNRISD.

Najam, A. (2001). The case against GEO, WEO or whatever-else- EO. In D. Brack & J. Hyvarinen (Eds.), *Global environmental institutions: Arguments for reform* (pp. 32–40). London: Royal Institute for International Affairs and Earthscan.

Norris, C. (2005). Partnerships for sustainable development the role of type II agreements. In A. Churie, G. Sjöstedt, & E. Corell (Eds.), *Global challenges furthering the multilateral process for sustainable development* (pp. 210–230). Sheffield: Greenleaf Publishing.

Oberthur, S., & Ott, H. E. (1999). *The Kyoto protocol international climate policy for the 21st century*. Berlin: Springer.

Olson, M. (1965). *The logic of collective action*. Cambridge: Harvard University Press.

Pianta, M. (2001). Parallel summits of global civil society. In H. Anheier, M. Glasius, & M. Kaldor (Eds.), *Global civil society 2001* (pp. 169–194). Oxford: Oxford University Press.

Redclift, M. (1987). *Sustainable development exploring the contradictions*. New York: Methuen.

Rees, W. (1990). The ecology of sustainable development. *The Ecologist, 20*(1), 18–23.

Reimann, K. D. (2006). A view from the top: International politics, norms and the worldwide growth of NGOs. *International Studies Quarterly, 50*(1), 45–67.

Sachs, O. (1993). *Global ecology: A new arena of political conflict*. London: Zed Books.

Schechter, M. G. (2005). *United Nations global conferences global institutions*. New York: Routledge.

Sheldon, D. (2000). Law, non-law and the problem of soft law. In D. Sheldon (Ed.), *Commitment and compliance: The role of non-binding norms in the international system* (pp. 4–10). Oxford: Oxford University Press.

Singh, K. (1994). *Managing common pool resources*. New York: Oxford University Press.

Stošić, B. D. (1964). *Les Organisations non Gouvernementales et les Nations Unies*. Geneva: Librairie Droz.

Strange, S. (1982). Cave! hic dragones: a Critique of Regime Analysis. *International Organization, 36*(2), 479–496.

Taylor, P. (1989). The origins and institutional setting of the UN special conferences. In P. Taylor & A. J. R. Groom (Eds.), *Global issues in the United Nations' framework* (pp. 7–34). New York: St. Martin's Press.

Tolba, M. (1998). *Global environmental diplomacy negotiating environmental agreements for the world, 1973–1992*. Cambridge: MIT Press.

UN. (1945). *Charter of the United Nations*. Retrieved from http://www.un.org/en/documents/charter

UN. (1947). *Yearbook of the United Nations 1946–1947*. New York: United Nations.

UN. (1968). *ECOSOC resolution 1296 (XLV)*. June 25, 1968.

UN. (1972). *General assembly resolution 2997 (XXVII)*. Retrieved from http://daccess-dds-ny.un.org/doc/RESOLUTION/GEN/NR0/270/27/IMG/NR027027.pdf?OpenElement

UNDPI. (2004). *NGOs and the United Nations Department of Public Information*. Retrieved from http://www.un.org/dpi/ngosection/brochure.htm

UNEP. (1982). *The state of the environment 1972–1982*. Nairobi: United Nations Environmental Program.

UNEP. (2001). *International environmental governance: Multilateral environmental agreements* (UNEP/IGM/1/INF/3).

von Moltke, K. (1996). Why UNEP matters. In H. O. Bergesen & G. Parmann (Eds.), *Green globe yearbook of international cooperation on environment and development* (pp. 55–64). Oxford: Oxford University Press.

Wagle, S. (1993). Sustainable development: Some interpretations, implications, and uses. *Bulletin of Science Technology Society, 13*(6), 314–323.

Walker, R. A. (2004). *Multilateral conferences purposeful international negotiation*. New York: Palgrave.

Wapner, P. (1996). *Environmental activism and world civic politics*. Albany: State University of New York Press.

WCED (World Commission on Environment and Development). (1987). *Our common future*. New York: Oxford University Press.

White, L. C. (1951). *International non-governmental organizations*. New Brunswick: Rutgers University Press.

Willetts, P. (1989). The pattern of conferences. In P. Taylor & A. J. R. Groom (Eds.), *Global issues in the United Nations' framework* (pp. 35–63). New York: St. Martin's Press.

Willetts, P. (1996a). Consultative status for NGOs at the United Nations. In P. Willetts (Ed.), *The conscience of the world* (pp. 31–62). Washington, DC: Brookings Institution.

Willetts, P. (1996b). Introduction. In P. Willetts (Ed.), *The conscience of the world* (pp. 1–13). Washington, DC: Brookings Institution.

Yamin, F., & Depledge, J. (2004). *The international climate change regime: A guide to rules, institutions and procedures*. Cambridge, UK: Cambridge University Press.

Young, O. (1997). *Global governance: Drawing insight from the environmental experience*. Cambridge: MIT Press.

Chapter 3
Theoretical Concepts

Abstract This chapter seeks to establish nongovernmental organizations influence on broader processes of global environmental governance. The chapter begins by establishing nongovernmental organizations' agency as one subcategory of global civil society. The next section reviews theoretical approaches to nongovernmental organizations' relationship with states and institutions. After determining that global civil society and regime theory do not match the boundary conditions of this case study, Sect. 3.4 reviews global environmental governance. Scholars within this field of study conclude that nongovernmental organizations may influence treaty outcomes at international environmental negotiations.

The remainder of the chapter incorporates time as a key dimension representing the broader processes of governance and establishes nongovernmental organizations' rational moral authority as a source of nongovernmental organizations' influence. Nongovernmental organizations convert their authority into influence by successfully achieving their goals. The section concludes by theorizing about factors which assist nongovernmental organizations in converting their authority into influence.

Keywords Nongovernmental Organizations (NGOs) • Influence • Rational moral authority • Global environmental governance • Global civil society • Regime theory

3.1 An Outline

This chapter begins by examining global civil society, regime theory, and global environmental governance. Scholars agree that environmental and developmental NGOs constitute a highly visible part of global civil society. Consequently, theories about global civil society establish NGOs as an important actor within international environmental politics.

A.E. Egelston, *Sustainable Development*, DOI 10.1007/978-94-007-4878-1_3,

The chapter continues by reviewing regime theory even though sustainable development cannot be considered a regime. While subdividing international environmental politics into constituent hard law treaties for the purpose of analysis has provided many fruitful avenues of scholarly research, it also leaves untouched other important areas of study. In short, areas of soft law should also be analyzed. That is not to say that all regime theoretical approaches should be rejected, they should not. Rather, theories designed to explain the relationship between states and NGOs may also be applicable to NGOs and other actors.

In recent years, global governance has joined regime theory as an important field of study. Global civil society and markets make important contributions towards providing predictability and stability to the ordering of global environmental governance. Global environmental governance provides a basis for combining global civil society's emphasis on non-state actors with regime theories' focus on states and institutions.

The second half of this chapter focuses on NGO influence, its sources, and its climb towards historical prominence. NGOs rational moral authority is a primary source of their influence and this influence permeates beyond formal negotiating sessions into the broader patterns of international affairs. Further, this section investigates NGOs ability to set and achieve goals that impact not only negotiating outcomes, but the structures, relationships, and normative values embedded within global environmental diplomacy.

3.2 Global Civil Society

Participants in the UN system consist of more than states and a brief review of any attendees list from a UN negotiating sessions will reveal a myriad of other actors ranging from scholars to businesses. These varying interests will have organized into an NGO for the purposes of receiving accreditation to the UN, but the reality of the situation is that the UN hosts a myriad of political actors beyond states. Nor is this situation unique to the UN as other groups such as the World Trade Organization (WTO), the North Atlantic Treaty Organization (NATO), and the Organization for Economic Co-Operation and Development (OECD) represent significant international political arenas. Each of these international structures has its own unique constituency, leading to an increase in both the number of platforms for interaction and the types of political actors that utilize these distinct political arenas. Morss (1991) theorizes that TNCs, international organizations and special interest groups joined states in providing order to the international system. In the not so distant past, scholars of international relations would ascribe power only to states. Despite the scholarly debate about the size and importance of these sources of power, the continued existence and activity of voluntary clubs and organizations seems to suggest that the international system seeks a fundamentally different means of existence beyond the state-centric bipolarity of the Cold War. Global civil society should not be seen as the smallest least powerful actor within international relations, but rather as an actor that extends well beyond the state-centric conceptualization of international organization.

Recent scholarship has led to an increasing number of terms seeking to describe these actors. These terms include private voluntary organizations (Korten 1990), pressure groups (Willetts 1982), transnational advocacy networks (Keck and Sikkink 1998), transnational civil society (Florini 2000; Smith et al. 1997), multipartite environmental governance (Meadowcroft 1999), global civil society (Anheier et al. 2001; Clark et al. 1998; Lipschutz 1992), and world civic politics (Wapner 1996).

Global civil society is a term of particular interest as it has multiple meanings and has been used to describe both an ideal type of actor as well as to refer to a singular, although complex, political actor. Clark et al. (1998:2) describe civil society as "frequent and dense exchange among individuals, groups and organizations in the public sphere separate from state-dominated action." Adding the term "global" to the definition of civil society suggests that the groups, individuals, and organizations are distributed geographically in every part of the earth. In this sense, global civil society is an ideal status for NGOs to achieve, rather than a specific categorization of actors. Only the largest NGOs achieve this global reach, and few ENGOs would make this claim of themselves.

In contrast, Wapner (1996:4) defines global civil society as "that domain of associational life situated above the individual and below the state." This definition includes families, churches and voluntary associations as well as political parties, NGOs, and labor groups, although certain scholars exclude markets from this definition (Cohen and Arato 1992:ix).[1] This definition of global civil society better represents the NGOs under investigation as part of the UN system. Formal participation within the UN does not require specific geographic reach, and NGOs may operate in as few as one country, or be organized in many countries. Further, NGOs authorized to participate at UN conferences are more likely to be in regular repeating contact with the UN bureaucracy than NGOs without formal UN recognition.

Scholte (1999:10) argued that global civil society "encompasses civic activity that (a) addresses transworld issues; (b) involves transborder communication; (c) has a global organization; (d) works on a premise of supraterritorial solidarity." He also points out that global civil society has both positive and negative implications. On the one hand, global civil society improves material well-being, provides additional civic education, promotes public participation, ensures adequate debate of public policy, increases transparency and accountability, promotes the legitimacy of governments, and improves social cohesion. However, they might also lack adequate resources, fail to deliver on promised goods and services, promote flawed policy, and suffer from a democratic deficit.

[1] Wapner's (1996:4) definition of global civil society allows for market and market proponents to be categorized as part of global civil society. However, the case studies utilized within this theory generally focus on non-market, non-state actors. Wapner's case study, for example, focused on Greenpeace, World Wildlife Fund and Friends of the Earth.

States interact with global civil society on two distinct levels. In one sense, states dictate the conditions under which global civil society participates in the political arena by the strictness with which they scrutinize groups located in their respective countries, the "top-down" approach. A "top-down" approach emphasizes structural approaches where established institutions promote or create NGOs for the purposes of supporting the established institution. For example, UNESCO deliberately promoted NGOs willing to support UNESCO's mission (Hoggart 1996:106). This approach also emphasizes the globalization of international political structures dominated by Western neo-liberal democratic values (Reimann 2006).

At the other end of the spectrum, global civil society has organized in locations where state control is ineffective. Global civil society uses this sheltered position to target changes in state behavior, also known as the "bottom-up" approach. Scholars studying the "bottom-up" approach to global civil society conclude that voluntary associations of people organized to impose moral and ethical limits on the uses of economic and political power have gained legitimacy as political actors (Cox 1999; Falk 1999). These groups deny states' presumptive role as the sole legitimate representation of its peoples' interests. Reasons for their formation include a decline in state power (Strange 1998), increased speed, reach, and reliability of communication (Matthews 1997:51) and specific local socio-economic conditions such as the state's inability to provide basic health and education services, particularly in the least developed countries (CIVICUS 1999).

A third viewpoint postulates that the prominence of global civil society is due in part to its capability to reconcile the top-down and bottom up approaches to environmental politics. That is, global civil society is most influential when it "links the local to international levels of politics" (Princen 1994: 33). Analysts working in this field cite the logic that grassroots organizations are most successful when they undertake a defined amount of work within the international system. That is, it is not enough for global civil society to exist. Rather it thrives when it seizes upon the political opportunity to expand its operations and to locate this new undertaking more broadly within the larger international system.

The current global civil society literature reflects the lack of a unified political philosophy about civil society. Perhaps unsurprisingly, theories about global civil society also lack unity and even a definitive agreement about basic characteristics. Consequently, scholars such as Anheier et al. (2001:3) speak about the necessity to map the contours of global civil society, while other scholars discuss its agency and influence. Regardless of how one views global civil society, these actors have become the focus of significant political importance as highlighted by their continued rapid growth, as well as the increasing willingness of older national and international governmental bodies to open previously closed decision-making processes (as in the case of the IMF and the World Bank). Additionally, global civil society increasingly serves as a significant conduit for both public and private development aid, particularly in the fields of health and education.

The dispute about the actors included within the contours of global civil society revolves around two dimensions: the inclusion of the neo-liberal capitalist market and the normative values and actions undertaken by global civil society. Typical academic definitions state that global civil society consists of non-state, non-market actors, while formal definitions of NGOs at the UN do not utilize this distinction, grouping all organizations, regardless of ideology, in a single, albeit diverse category. Scholars also exclude groups from global civil society based upon the normative belief of the organization. Consequently, organizations that do not have in mind the "common good" of mankind such as the Ku Klux Klan or the neo-Nazis have been eliminated from the definition of global civil society. Scholars such as Clark et al. (1998) promote the unity of normative values as a measure of the complete formation of a global civil society; that is civil society organizes around one singular set of universal values. This position makes for good theory, but in practice, does not exist. More problematically, the development of universal norms does not necessarily lead to either a unitary frame or a preferred political outcome. Measuring the unity of normative values to determine the completeness of global civil society suggests that the diversity and pluralism within global civil society will disappear over time. This homogeneity is fundamentally at odds with the diversity and institutional pluralism suggested by the description of global civil society (CIVICUS 1999).

The definitions of global civil society that place it below the state, above the individual and separate from the market are inadequate. These definitions invoke images of hierarchical control with states sitting at the top of the pecking order and individuals occupying the level of lowest political importance. Global civil society could be alternatively described as comprised of voluntary actors and organizations embedded within the state system, yet with resources and political influence that is not controlled by either states or markets. Consequently, global civil society can advocate altering perceived deficiencies within the socio-economic structure but nevertheless separate from state and market control. This definition of global civil society removes the language associated with direct hierarchical control, while retaining the socio-economic structures provided by states and markets. Perhaps the most succinct description of civil society was provided by Beem (1999:12) when he referred to civil society as "that part of society not under direct control of the state." Consequently, it would not be surprising to discover that countries with different state functions would have, at times, different conceptualizations of civil society.

Accordingly, scholarly research associates the increase of NGO participation with the end of the Cold War (Matthews 1997; Pianta 2001; Tracy 1994). According to this line of reasoning, the superpowers no longer needed the careful checks and balances that had come to symbolize international relations. This loosening of controls allowed the fragmentation of empires in both the East and the West, creating the political space for NGOs to organize. While this correlation of the end of the Cold War with the rise of NGOs undoubtedly holds true for much of Eastern Europe, Solidarity in Poland being a prominent example, NGOs in Western Europe and North America possess a longer pedigree.

Martens (2005) disagrees, arguing that what is new about the UN-NGO relationship is the increased opportunity to work with the UN. Her argument begins with the observation that NGOs form alliances with the UN, and international organizations more broadly, in order to achieve their policy goals. As a result of this alliance with the UN, NGOs increased their professionalism and bureaucratization, becoming more attractive to the UN. The UN responded by formally changing the rules and regulations governing NGO participation in order to increase the volume of NGOs in contact with the UN.

It seems likely that the sudden upswing of interest in civil society in the early 1990s stems directly from media coverage portraying Solidarity's success in reshaping the form of the Polish state.[2] Proponents of NGOs took advantage of this sudden breakthrough of political importance to advance their own standing within international politics, in essence claiming that NGOs are the organizations that aggregate into civil society. Ghils (1991) sought to describe the current state of affairs for NGOs by pointing out their increased proliferation. Their activities occur in three regions within the international state system. These three regions include areas where law does not yet exist, where public opinion believes states to be either morally lacking or illegitimate, and where states are either unable or unwilling to provide social order. Lipschutz (1992:393) adds that environmental politics is a political issue area where the activities of global civil society are "particularly visible." Wapner (1996) provides one of the first scholarly examinations of environmental organizations as part of world civic politics, a parallel forum for addressing global politics separate from the state system. He states that transnational environmental activist groups engage in an important work in creating environmental protection by focusing on areas where states have been unwilling or unable to go. It is not necessary that these activist groups are aware of their contributions to global civil society or of the quality of governance they provide. He adds to the theoretical literature by extending transnational environmental activist groups activities beyond a focus of attempting to influence states and details how these groups improve environmental protection within society as a whole.

Lipschutz (1996) adds to theories about global civil society by pointing out that organizations may work toward the changing of rules and of the processes of governance rather than towards creating new bureaucracies and structures. Consequently, global civil society has the potential to improve the quality of the environment because it can engage other actors at the local level. This engagement may include speaking directly with the actor responsible for environmental degradation or could include encouraging concerned individuals to voice their concern for the common good. The ability to influence other actors towards making environmentally-friendly decisions implies a diffusion of power, authority and legitimacy from states to non-state actors.

Global civil society adds to the understanding of NGOs because it highlights their separateness from states and businesses, especially TNCs. While these actors have sometimes adequately taken into account private individuals and their desires, states' need for survival along with businesses' preoccupation with profit are perceived as more legitimate concerns than that of the individual, or even of their

---

[2] Academic works that emphasize the role of civil society in the end of the Soviet Union include Cohen and Arato (1992) and Beem (1999).

collective. Global civil society allows the voice of the people to be heard separate from states and private businesses that have in the past, badly represented the interests of the citizen, particularly when normative topics are discussed such as the necessity for a healthy environment.

However, this body of work is too broad to be of help with an analysis of the political influence of NGOs in the history of sustainable development because it encompasses so many other actors that either are not interested in, or do not have access to, UN-sponsored activities. Global civil society aggregates into a different kind of actor, involved in a broader political arena than that of my case study. The literature comprising global civil society does include theories of interest about NGO influence, either as global civil society as a whole, or as a subset of a broader coalition of actors. These theories are reviewed in more detail in Sect. 3.5 NGO Influence.

3.3 Regime Theory

Ruggie (1972) published the seminal work in regime theory. He was concerned with two things. First, regime theory focuses on the conditions under which states collaborate with each other despite states' differing capabilities and differing objectives. Second, regime theory highlights the conditions under which such collaborative initiatives occurred. Ruggie utilized the term "regime", but failed to provide a definition to the term "regime". However, it is clear from the article that he intended regimes to focus on states' activities including negotiations and compliance organized around international treaty-making processes.

Regime theory gained proponents after Keohane and Nye's (1977) work in which they formulated the "complex interdependence" theory. However, regime theory would not become a major branch of international relations theory until 1982 when Stephen Krasner edited a special issue of *International Organization* focusing on regimes. Regimes can be defined as

> [I]implicit or explicit principles, norms, rules, and decision-making procedures around which actor expectations converge in a given area of international relations. Principles are beliefs of fact, causation, and rectitude. Norms are standards of behavior defined in terms of rights and obligations. Rules are specific prescriptions or proscriptions for action. Decision-making procedures are prevailing practices for making and implementing collective choice (Krasner 1982:186).

Regimes need not be legally recognized to exist. Norms and principles can be adhered to long before they become embedded within formal international law; in other words, specific international agreements need not have completed formal ratification. For example, the Law of the Sea constituted a regime well before its formal ratification 12 years after the end of the negotiating process (UN Office of Legal Affairs 2004). Additionally, regimes are not unique to environmental affairs, although investigation of environmental regimes is a primary focus of regime theorists.

The environment gained preeminence as a research area in international regime theory due to increasing concern about environmental problems, which in turn led to an increase in the number and prominence of environmental treaties. This increasingly dense issue, combined with an increase in scholarly research and analysis, led to concepts and ideas distinctly different from those that dominate traditional security and economic issues (Ferguson and Mansbach 2003). Environmental regime theory originated from attempts to extend principles originally derived from international political economy such as common property regime into environmental affairs to solve collective action problems (Young 1982, 1989b).

Certain branches of regime theory suggest that regimes are synonymous with institutions leading to the conflation of such terms as "institution" and "regime." Young (1999:7) begins by stating that regimes are lightly administered compared to organizations, leading to the necessity for actors within the regime to implement the regime. This lack of a regime bureaucracy forces the actors within the regime to deal with issues of implementation. Young next goes on to say that regimes' organizational capabilities increase over time as they acquire "offices, budgets, personnel, and legal personalities." At the same time, Young implies that the stabilization of the regime decreases the need for the actors to focus on implementation. This concurrence of regime and institution has led to a focus on institutional effectiveness as a key component of regime theory (Underdal 1992; Wettestad and Andresen 1991). Kütting (2000) challenges this focus, arguing instead that regimes should be evaluated in terms of their programmatic effectiveness, i.e. environmental effectiveness.

Within regime theory, it is not necessary that actor's expectations converge around changing *state* behavior, although this assumption dominates the field of environmental politics. A regime may form around converging expectations of *non-state actors* behavior. Businesses, for example, routinely adhere to regimes that regulate their internal affairs separate from governmental action, as voluntary regulation or self-regulation has grown dramatically over the last two decades. Examples of well-known self-regulatory environmental regimes include the ISO 14000 series as well as the Chemical Manufactures Association's Responsible Care Program.

Regime theory focuses on when, where, and why states undertake collective action within the international system instead of a solitary approach. Vig (1999:4) proposes four meaningful nuances of this definition of a regime. First, regimes involve specific issue areas. For example, the environment is not a regime, but ozone depletion is. Second, regimes emphasize cooperation over legal standing. For example, the Kyoto Protocol constituted a regime long before the treaty entered into force because it contained the norms underlying the climate change regime that detail reductions in greenhouse gas emissions. Third, regimes allow for non-state actors such as TNCs and NGOs to participate in both the formal negotiations and treaty implementation. Fourth, regimes evolve over time. Changes in agent and structure do not mean that the regime ends. Regimes terminate when the underlying philosophies and standards of behavior are no longer relevant and respected by multiple parties simultaneously. This distinction is separate from sporadic instances of regime non-compliance by an individual entity.

Regimes, as analytically constructed, provide structure and stability to the international system by encouraging cooperation over long time spans (Jervis 1982; Young 1989a). Regimes also constrain actors' behavior, frame issues to be addressed, and give a sense of order and predictability to international relations (Puchala and Hopkins 1982:246). Ideally, regimes allow for the successful construction of more efficient and cooperative governance. Within the environmental issue area, it is also desirable that regimes create more effective environmental protection (Kütting 2000), although this has not always been the case as certain regimes have failed to produce significant environmental improvement.

Regime theory with its focus on institutions and structures and the role of states within them is not entirely suitable for an analysis of sustainable development. Sustainable development is not a regime as there is no single corresponding institution housing all of the principles, rules, norms, and decision-making procedures for sustainable development. Further, there is no convergence of actor's expectations or behaviors that would lead to widespread cooperation around this issue.

Additionally, regime theory under-theorizes the role of NGOs relationships with other actors in the absence of a strong institution. In other words, regime theory focuses on a different actor, the state instead of NGOs. Where NGOs have been incorporated into regime theory, they alternatively act as a lobbyist to states participating in the formal negotiating processes, or as elements of a global society that bring legitimacy to the international system as a whole by representing people and their interests, rather than states (Breitmeier 2008). Despite these shortcomings, regime theory is important because it draws the boundaries of the theoretical analysis around a singular issue area and incorporates the correct groupings of actors. The downside of regime theory is that the actors' relationship with each other does not necessarily have explanatory usefulness.

3.4 Global Environmental Governance

Much of the scholarly literature included within the scope of global environmental governance revolves around the predominance of the Western neo-liberal capitalist socio-economic system that spread rapidly after the end of the Cold War, introducing a form of cultural hegemony.[3] Scholars additionally turned their attention to groups that have worked to resist these homogenizing forces during the same time – namely global civil society as so vividly portrayed by the Solidarity movement in

[3]Global environmental governance has also been used to discuss the need for a World Environmental Organization, particularly in the run-up to the World Summit on Sustainable Development in 2002 (Paterson et al. 2003:1). The topic continues to resurface among academic circles coinciding with each consecutive mega-conference, including the run-up to Rio+20.

Poland. At the same time that scholars recognized the increasing importance of global civil society, they realized that there was not necessarily a need to overturn regime theory in its entirety as many significant advances had occurred as part of this research agenda.

Underlying this inclusion of global civil society is the theoretical emphasis on the individual as a citizen instead of a mere voter or consumer. The citizen, then, is an individual agent who works together with others in a network for the promotion of the common good. Keck and Sikkink (1998) investigate one type of activist work, namely, transnational advocacy networks. This term is deliberately chosen to capture "what is unique about these transnational networks: they are organized to promote causes, principled ideas, and norms, and they often involve individuals advocating policy." The central position within these networks is occupied by international and domestic NGOs that work to coordinate discussion and diffusion of shared values and specialized knowledge.[4] The transnational advocacy network implicitly utilizes power because "networks influence politics at different levels (Keck and Sikkink 1998:8)," but there is no attempt to examine the causal linkages between knowledge, moral code, and power.

The transnational advocacy networks reported upon by Keck and Sikkink theoretically allow for BINGOs as long as they share the same values with the other members of the network. Despite Keck and Sikkink's identification of transnational advocacy networks as consisting mainly of NGOs, they never define NGO, a curious oversight given the highly contentious nature of the term. More likely, however, Keck and Sikkink envision BINGOs as sharing a unique network separate from the transnational advocacy network, as they disallow groups from joining with others for the wrong reasons. In reality, businesses may join a network because it is the most profitable for their company without sharing values with the advocacy network. Keck and Sikkink do not allow actors to do the right things for the wrong reasons, although they do recognize leverage as a means to persuade others to change behaviors.

Further, grouping all NGOs into one transnational advocacy network oversimplifies the plurality of NGO beliefs and values. While most environmental groups advocate for increased environmental protection, there is a wide variance in beliefs about how to accomplish this task. In other words, some NGOs are "more green" then others. At one end of the spectrum conservationist organizations focus on procuring land in order to protect the natural habitat through good stewardship. Organizations like Ducks Unlimited will work towards good stewardship by purchasing and managing wetlands to preserve the environment for future use, including recreational hunting. On the other end of the spectrum, Greenpeace will invoke the wrath of governments in order to achieve environmental protection.

[4] While it is a premise of this text that networks are, normatively speaking "good", it should be acknowledged that criminal networks also affect global environmental governance. Duffy (2005) details how the "shadow state" inhibits global environmental governance in Madagascar.

Greenpeace has a long track record of boarding privately held and government sanctioned vessels in order to prevent activities ranging from oil drilling to whaling.

Rather than assuming that NGOs coalesce into one transnational advocacy network, this research views NGOs as multiple transnational advocacy networks, defined around their underlying beliefs and values. For the purposes of evaluating NGO activity around sustainable development, networks can be identified based upon their location around these discourses. For example, a transnational advocacy network exists focusing on the limits discourse. A separate transnational network organized around continued economic growth. Further, membership in one network does not preclude membership in another differing network as long as the differences in values can be resolved. On a related note, the composition of the networks may also fluctuate over time. This may be particularly relevant if differences in values between the competing networks cannot be resolved.

More recently, scholarly debate has focused on how these networks create social change, providing what Roseneau and Czempiel (1992) called "governance without government." The authors distinguish between government (an organization) and governance (a process). Authors such as Princen and Finger (1994) focus on NGOs' capability to link the local with the global, a necessity within international environmental politics caused by the unparalleled shift in power and emphasis away from the national to both the global and the local.[5] For these authors, NGOs engage in social learning as well as provide specialized knowledge about environmental problems which in turn leads to challenging both the control and the legitimacy of the state. Kütting and Lipschutz (2009) investigate the relationship between power and knowledge, and space and place. They argue that knowledge must be understood in light of its place of origin. Universal environmental knowledge at the global level can be in conflict with specific knowledge at the local level.

Raustiala (1997:727) criticizes Princen and Finger's "local–global" theory that describes NGOs rather than produces explanatory theory. Instead, he believes that states take advantage of NGOs' specialized knowledge to create a "pool" of policy ideas, essentially outsourcing a primary expense in the negotiation process. The more "free" policy ideas the state receives from NGOs, the more resources the state retains to spend elsewhere. Thus for Raustiala, NGOs enhance the prestige of the state at the international negotiating table precisely because of the state's intermediary role between widely varying elements of its domestic society. The corollary to Raustiala's theory is that "free" NGO services have more value for resource-poor countries, assuming that NGO's proposed policies support their position. For example Vanuatu, a small island state, allowed an NGO to represent it at the climate change negotiations.

[5] Wapner (1996:8–9) used the concepts of sub-statism and supra-statism to describe this phenomenon.

To date, scholars have not given much thought to global civil society's capacity for innovation and experimentation in generating new public policies, although it has been hinted at by scholars such as Sanwal (2004:20) when he states that:

> A new approach towards international cooperation is emerging that combines the knowledge of practitioners with an emphasis on innovation and exchange of experiences, to take advantage of the opportunities provided by increasing flows of trade, investment, and technologies to promote the transition to sustainable development.

Unfortunately, Sanwal fails to follow through on this recognition of the need for innovation and multiplication of policy programs. Further, Sanwal unnecessarily limits the opportunities for such innovation and experimentation to the advantages of trade, investment, and technology. Opportunities to innovate and experiment are more broadly generated and the sources of innovation and experimentation include structural opportunities and brainstorming sessions by transnational activist networks, particularly where normative values enter into the political realm. Consequently, a diverse global civil society contributes to policy innovation and experimentation by providing more locations where these novel ideas and policies can be examined safely. Lastly, Sanwal states that innovation should occur within institutions, while at the same time acknowledging relationships among states, international regimes, and non-state actors. Given that these three actors are indeed in relationships, it is not necessarily true that institutions are, or should be, the only location for innovation and experimentation. That is, "good ideas" generated outside of the institution are easily transferable back to the institution, particularly where well-defined pre-existing relationships have been established.

This function of innovation and experimentation is important for four reasons. First, it defines the boundaries of current negotiations at the UN mega-conferences. Policy options have been both brought to and removed from the negotiating table based upon the results from trial policy programs. A particularly noteworthy example comes from the climate change regime when the computer tracking system from the United States acid rain trading system was modified for tracking greenhouse gas emissions. Consequently, the design of the mechanism by which the carbon credits could be traded occurred before the finalization of the rules governing the Kyoto Mechanisms. Second, pilot programs are not necessarily easy to end as programs can be indefinitely extended or become embedded within formal treaties. Third, innovation and experimentation within global civil society are necessary to overcome the inherent inertia within the socio-economic system. As students of global environmental politics correctly point out, cultural preferences can override environmental priorities, as the case of whaling in Japan and Norway clearly demonstrates. Lastly, in the absence of hard law, policy experiments serve as a de facto governance system regardless of whether the program is a complete disaster or much heralded success. This feature would suggest that implementation occurs separately from the indirect method of national regulations. For environmental policy, this conclusion intuitively makes sense as many states function as a source of indirect pressure, regulating the production of goods and services rather than producing commodities directly.[6]

Consequently, one result of the increased numbers of NGOs is the corresponding increase in trial policy programs in place. These trial programs take place for two interrelated reasons – the increasingly complex environmental problems such as protection of biodiversity and climate change, as well as the increased complexity within the contemporary economic system, frequently referred to as globalization. The "newness" of these two reinforcing issue areas raises the need for novel regulations and policies. As these trial programs overlap each other, they give a definitive sense of order to global environmental governance. Further, these programs become acceptable policy options in the absence of state consensus.

The concepts of innovation and experimentation differ from traditional forms of lobbying such as providing expert information, legal opinions, and new public policy such as one might expect from a "think tank." The innovation and experimentation as used here is both broader and deeper than these traditional functions. There is a subtle but important difference between generating the idea for a new policy, and actively engaging in trial programs to demonstrate new policies. Generating an idea for a new public policy might well be construed as lobbying, while orchestration and/or participation in a pilot program moves well beyond a lobbying stance into governance.

Lafferty and Meadowcroft (2000:379) also focus on innovation and experimentation. However, they do so in the context of government/business relationships.

> Most governments (regardless of ideological persuasion) seem to have acknowledged that relationships with business are crucial to the realization of innovative practices, so that it is hardly surprising that officials have gone out of their way to explain that sustainable development is compatible with economic growth and successful business.

Lafferty and Meadowcroft go on to describe the dynamics necessary for policy innovation: opportunity, responsibility, and partnership. This same characterization also applies to NGOs contribution to global environmental governance.

To summarize, global environmental governance contributes to the model for this research because it highlights NGOs' role of promoting social change by working within the system to promote new normative standards. This theoretical approach thus involves the correct actor, performing a similar set of roles. NGOs and the networks they create have actively promoted sustainable development, along with many other beneficial ideas in the fields of environment, development and human rights. Further, NGOs also engage in other activities that are closely associated with normative standards including innovation and experimentation with new public policies.

---

[6]That is not to say that government bureaucracies do not cause environmental degradation as militaries frequently contribute to pollution problems. Additionally, governments leave their own environmental footprint along with the activities they regulate.

3.5 NGO Influence

The previous sections reviewed two common strands of international relations theory – global civil society and global environmental governance. Neither field precisely represents the case study of this research. Global civil society defines the political arena too broadly. Global civil society incorporates too many heterogeneous actors into this category, while defining NGOs as only those organizations that formally participate in regime negotiations is too restrictive. Global environmental governance more closely matches the theoretical structure for this research; however, past studies limit NGOs influence to a knowledge provider.

That is not to say that individual theories within this field have nothing instructive to say about NGOs desire to influence politics at the UN. There is a substantial literature upon the role and effects of NGOs within international relations; much of this scholarly literature assesses NGO attempts to influence states. However, there are other agents within international relations that NGOs may target, including international organizations as well as other NGOs. In other words, NGOs do not focus their efforts solely on states, but also on other actors within global environmental politics more broadly. Similarly, NGOs also seek to influence not only formal treaty negotiations, but also so-called "soft law" documents such as declarations, reports, and voluntary commitments. Additionally, NGOs may also seek to influence the processes of governance in order to secure beneficial procedural rules and permanent access to not only UN staff but also access to informal hallways and miscellaneous meetings. NGOs do so in order to improve their status and acceptability to other actors.

3.5.1 Rational Moral Authority

NGOs use their authority as rational moral actors to advance their goals by influencing actors, structures, negotiations, and other documents. Thomas (2007:95) posits that international NGOs intrinsically possess rational moral authority due to their promotion of universal human interests, support of democratic goals and organizations, and commitment to global progress. NGOs seek, through political processes, to change the behavior of other actors including states, international organizations, economic powerhouses, and other entities with differing moral values. Further, NGOs may also seek to change the prioritization of values in cases where underlying norms and principles are shared, but ascribed differing hierarchy.

As a rational moral agent, NGOs act in their own self-interest in support of an overarching principled idea or normative value. Rational-moral agency implies a hierarchy of values and preferences pursued in a logical sequence, given the constraints of time, financial resources, and scientific understanding. Rational moral agency further assumes that all people are, or should be, free and equal. Rational moral agency does not imply shared hierarchies of values and preferences across all rational moral agents. The hierarchy of normative values varies from agent to agent. Additionally, they are not static in that the same individual agent may change their own ranking of normative values over time.

3.5.2 Goal Achievement

Rational moral agents pursue action based upon their goals; identification of goals allows insight into the means used to obtain these goals. According to Feraru (1974), NGOs changed mentality from an "apolitical" actor who provided specialized knowledge to decision makers to a "political" actor who expressed preferred outcomes. This change occurred around the time of the Stockholm conference in 1972. At that time NGOs demonstrated the ability to engage in activities to further their own agenda. Feraru (1974:49) suggested NGOs pursued four key goals at the Stockholm conference – (1) getting information about the environment and about potential solutions for their own use, (2) giving expert assistance to member states and UN officials, (3) lobbying other actors to consider specific policy principles and programs, and (4) representing the organization during formal negotiations.

Feraru derived these four goals from surveying NGOs that participated in the Stockholm conference. As such, these goals were intended to capture NGOs activities at a specific moment in time, namely the UNCHE conference in 1972. She never intended these goals to represent NGO activities over a longer period of time. To compensate for the differences in time scale, the definitions of these four goals should also be updated. Additionally, Feraru assumed that NGOs focused their activities to influence government officials (states) or UN bureaucrats. This reformulation widens the targets of these activities to include other actors, including, but not limited to NGOs and UN staff.

Feraru's first goal, getting information, involves collecting data "about the global environment to transmit to the organization's members or to the wider public." She also places mobilizing public support (or criticism) for Stockholm's outcomes in this category. While mobilizing public support can reasonably be included with representing the organization, mobilizing support is included with the data collection efforts because it allows for an analysis of NGOs rational moral authority while transforming data into normative-based action. Collecting information about proposed political solutions is added to this definition in order to capture the continuity of meetings that form the backdrop of the emerging sustainable development discourse.

Feraru's second goal, giving expert assistance, focuses on NGOs role as a provider of specialized knowledge and expertise to decision-makers. She assumes that NGOs will continue to give expert assistance past the Stockholm conference and she incorporates implementation of the Stockholm Action Plan into this category as well. Instead, these activities are split into two goals. Giving expert assistance is defined as the provision of specialized knowledge and expertise to decision-makers. However, grouping NGOs implementation of the Stockholm

Action Plan with the giving of expert assistance obscures a vitally important facet of global environmental governance, the implementation of the policy outcomes. Consequently, policy implementation is separated into its own goal.

Feraru (1974:40) narrowly defined her third goal, lobbying decision-makers, as "trying to persuade UN bodies or governments to consider or adopt some principle or program that the INGO [international NGO] supports." This research uses a broader definition by adding other actors to the group of players that NGOs may lobby to adopt the norms of sustainable development discourse. This definition more closely aligns with theoretical approaches within global environmental governance that accepts global civil society as important actors.

Feraru's fourth goal, representing the organization, is defined as "expressing to UN officials and government representatives the general outlook of the organization on environmental matters." She goes on to state that representation was less likely to be cited as an NGO objective after Stockholm and speculates this could be due to a more precise definition of objectives after participating in the Stockholm conference. Instead, this research theorizes that NGOs became sensitized to the peculiarities of UN politics, namely, that states, and states alone, are the sole legitimate representatives of their citizens. NGOs who were restricted on paper, but not in practice, to providing expert assistance and to mobilizing public support for conference outcomes, tread lightly around this expectation lest they lose their newfound access to negotiations.

Despite the lack of support for representing the organization in Feraru's original survey, this remains an important goal of NGOs, particularly over longer time frames. Like the other goals derived from Feraru's work, this research uses a broader definition to incorporate a wider audience as the target of NGOs' activities. Accordingly, representing the organization means that NGOs express the outlook of the organization, by communicating the norms, beliefs and values to other political participants.

Previously, this research added policy implementation to the list of goals NGOs seek to achieve. This activity is an understudied aspect of NGOs contributions to international relations. Policy implementation represents not only the outcomes, recommendations and requirements of formal treaty negotiations, but also policy initiatives stemming directly from NGOs rational moral agency such as creating voluntary initiatives for the purposes of obtaining environmental protection. Further, NGOs gain valuable insight into environmental affairs through an increase not only in their knowledge base and public visibility, but also through the material resources expended during implementation.

Caldwell (1988:19) adds that one of the functions of NGOs within international politics has been to provide continuity in international environmental politics during the periods between formal negotiating conferences. That is, NGOs help to keep environmental negotiations "alive" by putting pressure on governments, international organizations, and BINGOs not only at negotiating sessions, but also during the long breaks in between. NGOs' long-term support for environmental protection probably resulted in significant improvement to the environment and to NGOs' improved political capabilities. NGOs' pressure should not be narrowly construed as merely lobbying for national legislation on behalf of the environment, although this function is one of the more important activities that NGOs undertake. NGO continuity also entails such functions as gathering information about the state of the environment and about others' attitudes and their likelihood to act in an environmentally friendly manner in the future – information that can be utilized in the process of creating international environmental governance.

3.5.3 Political Influence

Achieving any (or all) of these goals leads to political influence. Cox and Jacobson (1973:3) provide a generic definition of influence when they state "influence means the modification of one actor's behavior by that of another." For the purposes of this project, the actor is presumed to be a political decision-maker. This political decision-maker could vary by both political arena and time, thus, it becomes necessary to identify the political decision-maker(s) NGOs target at various points in the analysis such as states, intergovernmental organizations, or other NGOs.[7]

Arts (1998: 58) provides more specificity when he articulates influence as "the achievement of (a part of one's) policy goal with regard to an outcome in treaty formation and implementation which is (at least partly) caused by one's own and intentional intervention in the political arena and process concerned." Rather than link influence to the achievement of a policy goal, this research links influence to goal achievement more broadly. NGOs may meaningfully impact global environmental governance by achieving their goals with respect to processes or the spreading of norms, separate from policy outcomes.

Betsill and Corell (2008:24) further restrict influence to "when one actor intentionally communicates to another so as to alter the latter's behavior from what would have otherwise occurred." This restricts sources of NGO influence to instances where communication leads to goal achievement in an international environmental negotiation. It also neglects important facets of NGO resources – in particular, their ability to use their specialized knowledge as a substitute for funding in order to embed norms into documents. However, the idea that influence represents a player's successful attainment of goals is sufficiently broad to accommodate influence through the use of normative beliefs, financial resources, and expert knowledge.

The possibility that NGOs may achieve their goals, but not obtain the expected outcome also exists. There is a possibility of unintended consequences. Merton

[7]The book assumes that intergovernmental organizations are not controlled by states, but rather than bureaucrats act as rational legal authorities. Barnett and Finnemore (2004:31) posit that "IOs are able to use their authority, knowledge, and rules to regulate and constitute the world that subsequently requires regulation. ... IOs (1) classify the world, creating categories, of problems, actors, and action; (2) fix meanings in the social world; and (3) articulate and diffuse new norms and rules."

(1936) examined the unintended consequences of public policy. He classified unintended consequences into three categories – unexpected benefits, unexpected drawbacks and perverse results. These outcomes apply to political influence as well as to public policy. Arts (1998) disregarded the impacts of unintended consequences due to the difficulty in assessing the impacts. This book argues that unintended consequences likely altered NGOs standing within the international community, and will, if possible, analyze the effects of the unintended consequences on NGOs political influence.

It is also possible that NGOs target the political decision-maker indirectly (Potter 1996:43) by targeting an intermediary organization. For example, organization A may influence organization B to speak to organization C for a myriad of reasons, including organization's A judgment that organization C will respond more favorably if a different messenger communicates organization A′s request. Influence cannot be quantitatively measured. It is possible to count the number of times interactions between groups occurred, but it is not nearly so straightforward to determine the results of this interaction, i.e., to determine the extent to which the recipient's behavior was altered, or to determine conclusively causality. Because of these limitations, it is necessary to speak of indications of influence, rather than to count number of participants, or length of presentations, although these things are not unimportant in and of themselves. That is not to say that identifying and categorizing political influence is not possible, but rather that evidence of influence should link NGO activities to goal attainment across a broad spectrum of activities ranging from agenda setting to treaty implementation. Betsill and Corell (2008:38) suggest that scholars "assess" levels of NGO influence into three categories – low, moderate and high levels of influence – based upon whether or not NGO participation affects either the negotiating process or outcomes. Similarly, this book also utilize this schema; however, classifications of levels of influence vary slightly from Betsill and Corell's due to the differences in our definitions and case studies. Accordingly, this book define a low level of NGO influence as achieving a goal, but no changes in norms or outcomes such as policies, structures, or procedures. Medium NGO influence represents cases where NGOs achieve their goals with some visible changes in norms and outcomes. High NGO influence represents cases where NGOs achieve their goals with some visible changes in norms and outcomes directly attributable to NGO activities.

3.5.4 Power and Influence

Scholars debate the relationship between power and influence. Dahl (1957) uses power interchangeably with influence. In this case, a finding of NGO influence would suggest that NGOs also have power. Such a finding seemingly contradicts NGOs admittedly fragile grasp when participating at the UN. States can and have, successfully removed NGOs from the negotiating space, although there are increasingly negative ramifications for exercising this option.

Power and influence do not describe the same political phenomenon. Cox and Jacobson (1973:3) distinguish between power and influence by defining power in terms of capabilities, or "the aggregate of political resources that are available to an actor." This definition also does not adequately address the differences between power and resources as the largest and most wealthy NGOs may have more financial resources and specialized knowledge than many small states and/or international organizations. Nye (1990) notes that power is reduced to resources in order to make power easy to measure. In other words, power reflects the capability of an agent to achieve their own goals using both physical resources such as labor and raw materials, and normative resources such as beliefs and knowledge.

Arts (1998) defined political power as "one's more or less permanent ability to influence policy outcomes." Earlier in this section, this book rejected linking influence to policy outcomes. Similarly, it also rejects the notion that power should be linked to the permanent ability to influence policy outcomes. However, Arts' definition of power incorporates a key element of this study – that is, time. Consequently, this book uses an admittedly simplistic view of power; that is, power can be defined as the consistent attainment of influence over time.

3.5.5 Explanatory Factors

Explanatory factors are descriptors that help identify under what conditions NGOs may be more or less likely to have influence. Explanatory factors describe the conditions or the tools that help NGOs achieve their goals. Arts (1998) and Betsill and Corell (2008) derive numerous explanatory factors for NGO influence during international environmental negotiations. As discussed previously, these factors may not be an appropriate framework for this research as their political arena differs from this work. The principles that Arts and Betsill and Corell used to derive these explanatory factors differ significantly from the broader definitions utilized in this case study. Consequently, the conditioning factors used by both Arts and Betsill and Corell are similarly too narrow for use in this research. Five new explanatory factors emerge based upon the broader theoretical approaches outlined above including (1) normative traits and characteristics, (2) NGO capabilities, (3) political circumstances, (4) players' expectations, and (5) structural rules of engagement. This framework will be utilized to evaluate under what conditions NGOs were more likely to achieve their goals, and influence the sustainable development discourse.

3.6 A Model Emerges

Each of the bodies of work makes an important contribution in part to the theoretical model utilized in the empirical analysis of this research, although none of the above are adopted in its entirety. The first two bodies of literature, global civil society and regime theory, are relatively straightforward. Global civil society includes a non-state actor acting on behalf of concerned individuals in favor of a normative standard; i.e., sustainable development, that continues to be underemphasized by states and markets. Regime theory contributes a focus on structures and institutions as negotiated through the formal meetings of states and other observers, in this case the UN, its specialized agencies, particularly UNEP, and the first environmental mega-conference overseen by the UN General Assembly.

Global environmental governance emphasizes the role of NGOs in norm development and specialized knowledge, both within and outside of the UN and with respect to a variety of other actors including states, TNCs, international governmental organizations such as the World Trade Organization, and other NGOs. With respect to this research, it is necessary to expand NGOs' activities into areas other than norm development and specialized knowledge into the roles of project implementation and consolidation, two areas that global environmental governance has yet to investigate.

Theorizing that NGOs are rational moral agents operating within a political arena allows the linkage between the character of NGOs with their goals in order to analyze NGO influence. As rational moral actors, NGOs pursue survival of the organization simultaneously with expansion of the primary normative discourse. Like all other organizations, survival trumps expansion of the primary normative discourse. Without the organization, the structures promoting the normative discourse collapse. Defeat the primary normative discourse, however, and the organization is likely to adopt a slightly different yet similar discourse and continue advocating the new discourse.

These six primary goals – (1) getting information, (2) providing expert assistance, (3) lobbying policymakers about specific policies and preferred outcomes, (4) representing the organizations values and norms to other players, including policymakers, (5) implementing policy, and (6) keeping the agenda visible, stem from the rational moral character of NGOs. Goal 1 derives from survival of the organization, while goals 2, 3, and 6 derive from expansion of the primary normative discourse. Goals 4 and 5 stem from both the survival of the organization and from expansion of the primary normative discourse because the two goals are intertwined with the *raison d'être*. NGOs exercise influence when they achieve these six goals.

References

Anheier, H. K., Glasius, M., & Kaldor, M. (2001). *Global civil society 2001*. Oxford: Oxford University Press.

Arts, B. (1998). *The political influence of global NGOs: Case studies on the climate and biodiversity conventions*. Utrecht: International Books.

Barnett, M., & Finnemore, M. (2004). *Rules for the world: International organizations in global politics*. Ithaca: Cornell University.

Beem, C. (1999). *The necessity of politics reclaiming American public life*. Chicago: The University of Chicago Press.

Betsill, M., & Corell, E. (2008). *NGO diplomacy: The influence of nongovernmental organizations in international environmental negotiations*. Cambridge: The MIT Press.
Breitmeier, H. (2008). *The legitimacy of international regimes*. Burlington: Ashgate.
Caldwell, L. K. (1988). Beyond environmental diplomacy: The changing institutional structure of international cooperation. In J. E. Carroll (Ed.), *International environmental diplomacy the management and resolution of transfrontier environmental problems* (pp. 13–28). Cambridge: Cambridge University Press.
CIVICUS. (1999). *Civil society at the millennium*. West Hartford: Kumarian Press.
Clark, A. M., Friedman, E. J., & Hochstetler, K. (1998). The sovereign limits of global civil society: A comparison of NGO participation in UN world conferences on the environment, human rights and women. *World Politics, 51*(1), 1–35.
Cohen, J. L., & Arato, A. (1992). *Civil society and political theory*. Cambridge: MIT Press.
Cox, R. W. (1999). Civil society at the turn of the millennium: Prospects for an alternative world order. *Review of International Studies, 25*(1), 3–28.
Cox, R. W., & Jacobson, H. K. (1973). *The anatomy of influence: Decision making in international organizations*. New Haven: Yale University Press.
Dahl, R. (1957). The concept of power. *Behavioral Science, 2*(3), 201–215.
Duffy, R. (2005). Global environmental governance and the challenge of shadow states: The impact of illicit sapphire mining in Madagascar. *Development and Change, 36*(5), 825–843.
Falk, R. (1999). *Predatory globalization: A critique*. Cambridge: Polity Press.
Feraru, A. T. (1974). Transnational political interests and the global environment. *International Organization, 28*(1), 31–60.
Ferguson, Y. H., & Mansbach, R. W. (2003). *The elusive quest continues theory and global politics*. Upper Saddle River: Prentice Hall.
Florini, A. M. (2000). *The third force: The rise of transnational civil society*. Washington, DC: Carnegie Endowment.
Ghils, P. (1991). International civil society: International non-governmental organizations in the international system. *International Social Science Journal, 133*(3), 417–429.
Hoggart, R. (1996). UNESCO and NGOs: A memoir. In P. Willetts (Ed.), *The conscience of the world* (pp. 98–115). Washington, DC: The Brookings Institution.
Jervis, R. (1982). Security regimes. *International Organization, 36*(2), 357–378.
Keck, M. E., & Sikkink, K. (1998). *Activists beyond borders: Advocacy networks in international politics*. Ithaca: Cornell University Press.
Keohane, R. O., & Nye, J. S. (1977). *Power and interdependence: World politics in transition*. Boston: Little, Brown.
Korten, D. C. (1990). *Getting to the 21st century voluntary action and the global agenda*. West Hartford: Kumarian Press.
Krasner, S. D. (1982). Structural causes and regime consequences: Regimes as intervening variables. *International Organization, 36*(2), 185–205.
Kütting, G. (2000). *Environment, society and international relations. Towards more effective international agreements*. New York: Routledge.
Kütting, G., & Lipschutz, R. (2009). *Environmental governance: Power and knowledge in a local–global world*. London: Routledge.
Lafferty, W. M., & Meadowcroft, J. (2000). *Implementing sustainable development strategies and initiatives in high consumption societies*. Oxford: Oxford University Press.
Lipschutz, R. D. (1992). Reconstructing world politics: The emergence of global civil society. *Millennium, 21*(3), 389–420.
Lipschutz, R. D. (1996). *Global civil society and global environmental governance: The politics of nature from place to planet*. Albany: State University of New York Press.
Martens, K. (2005). *NGO's and the United Nations: Institutionalization, professionalization and adaptation*. Basingstoke: Palgrave Macmillan.
Matthews, J. (1997). Power shift. *Foreign Affairs, 51*(1), 50–66.
Meadowcroft, J. (1999). The politics of sustainable development emergent arenas and challenges for political science. *International Political Science Review, 20*(2), 219–237.

Merton, R. K. (1936). The unanticipated consequences of purposive social action. *American Sociological Review, 1*(6), 894–904.
Morss, E. (1991). The new global players: How they compete and collaborate. *World Development, 19*(1), 55–64.
Nye, J. (1990). *Bound to lead: The changing nature of American power*. New York: Basic Books.
Paterson, M., Humphreys, D., & Pettiford, L. (2003). Conceptualizing global environmental governance: From interstate regimes to counter-hegemonic struggles. *Global Environmental Politics, 3*(2), 1–10.
Pianta, M. (2001). Parallel summits of global civil society. In H. Anheier, M. Glasius, & M. Kaldor (Eds.), *Global civil society 2001* (pp. 169–194). Oxford: Oxford University Press.
Potter, D. (1996). Non-governmental organizations and environmental policies. In A. Blowers & P. Glasbergen (Eds.), *Environmental policy in an international context III. Propects for environmental change* (pp. 25–50). London: Arnold.
Princen, T. (1994). NGOs: Creating a niche in environmental diplomacy. In T. Princen & M. Finger (Eds.), *Environmental NGOs in world politics: Linking the local and the global* (pp. 29–47). London: Routledge.
Princen, T., & Finger, M. (1994). *Environmental NGOs in world politics: Linking the local and the global*. London: Routledge.
Puchala, D. J., & Hopkins, R. F. (1982). International regimes: Lessons from inductive analysis. *International Organization, 36*(2), 245–275.
Raustiala, K. (1997). States, NGOs and international environmental institutions. *International Studies Quarterly, 41*(4), 719–740.
Reimann, K. D. (2006). A view from the top: International politics, norms and the worldwide growth of NGOs. *International Studies Quarterly, 50*(1), 45–67.
Roseneau, J., & Czempiel, E.-O. (1992). *Governance without government: Order and change in world politics*. New York: Cambridge University Press.
Ruggie, J. G. (1972). Collective goods and future international collaboration. *American Political Science Review, 66*(3), 874–893.
Sanwal, M. (2004). Trends in global environmental governance: The emergence of a mutual supportiveness approach to achieve sustainable development. *Global Environmental Politics, 4*(4), 16–22.
Scholte, J. A. (1999). *Global civil society: Changing the world* (Working Paper No. 31/99). Coventry: Center for the Study of Globalization and Regionalization.
Smith, J., Chatfield, C., & Pagnucco, R. (1997). *Transnational social movements and world politics: Solidarity beyond the state*. Syracuse: Syracuse University Press.
Strange, S. (1998). *Mad money when markets outgrow governments*. Ann Arbor: The University of Michigan Press.
Thomas, G. (2007). Globalization: The major players. In G. Ritzer (Ed.), *The Blackwell companion to globalization* (pp. 84–102). Oxford: Blackwell.
Tracy, C. (1994). The roots of influence: NGOs and the relationship between human rights and the environment. *Journal of International Law and Practice, 3*(1), 21–46.
Underdal, A. (1992). The concept of regime effectiveness. *Cooperation and Conflict, 27*(3), 227–240.
UN Office of Legal Affairs, Division for Ocean Affairs and the Law of the Sea. (2004). *United Nations convention on the law of the sea. A historical perspective*. Retrieved from http://www.un.org/Depts/los/convention_agreements/convention_historical_perspective.htm#Historical%20Perspective
Vig, N. J. (1999). Introduction: Governing the international environment. In N. J. Vig & R. S. Axelrod (Eds.), *The global environment institutions, law and policy* (pp. 1–26). Washington, DC: CQ Press.
Wapner, P. (1996). *Environmental activism and world civic politics*. Albany: State University of New York Press.
Wettestad, J., & Andresen, S. (1991). *The effectiveness of international resource cooperation, some preliminary findings* (R:007). Oslon: The Fridtjof Nansen Institute.

Willetts, P. (1982). *Pressure groups in the global system. The transnational relations of issue-orientated non-governmental organizations*. New York: St. Martin's Press.
Young, O. (1982). *Resource regimes: Natural resources and social institutions*. Berkeley: University of California Press.
Young, O. (1989a). Science and social institutions: Lessons for international resource regimes. In S. Andersen & W. Ostreng (Eds.), *International resource management: The role of science and politics* (pp. 7–25). New York: Belhaven Press.
Young, O. (1989b). *International cooperation building regimes for natural resources and the environment*. Ithaca: Cornell University Press.
Young, O. (1999). *Governance in world affairs*. Ithaca: Cornell University Press.

Chapter 4
From Stockholm to Our Common Future

Abstract This chapter presents the historical narrative for this project; namely, the political history of the sustainable development discourse. The purpose of this chapter is to present a cohesive story about the actors, agents, and structures that overcame problems stemming from the East-West conflict, the North-South gap, and existing rules and traditions of state diplomacy to create a new environmental paradigm. Although sustainable development may not harden into a regime, it nevertheless permanently altered the relationship not only between man and his environment, but also between each other.

The chapter also reviews how the limits discourse shaped one conceptualization of sustainability. This chapter contains two distinct discourses dealing with sustainable development. The first discourse was promoted by the United Nations and included an emphasis on continuing economic growth. The Club of Rome funded initial research on a second discourse emphasizing a steady state economy around finite limits. George Mitchell, owner of Mitchell Energy and a prominent member of the Club of Rome continued funding and promoting this limits discourse.

Keywords United Nations Conference on Human Environment (UNCHE) • United Nations Environment Protection (UNEP) • Nongovernmental Organizations (NGOs) • Maurice Strong • World Commission on Environment and Development (WCED)

4.1 Environmental Affairs Prior to Stockholm

Environmental issues played a limited role in UN affairs in the first 20 years of its existence. However, UNESCO, a specialized agency within the UN "family" whose function is self-described as "to build peace in the minds of men" (UNESCO 2005), made two contributions to the early institutionalization of international

A.E. Egelston, *Sustainable Development*, DOI 10.1007/978-94-007-4878-1_4,

environmental politics that merit brief attention.[1] First, UNESCO, in 1948, played a pivotal role in the formation of the IUCN. Julian Huxley, the first Secretary-General of UNESCO, worked tirelessly for the establishment of NGOs to fill perceived gaps within international civil society. In particular, IUCN, as its name implies, worked towards the protection and conservation of wildlife and natural habitats. This quasi-governmental organization, comprising both governments and environmentally friendly NGOs, historically enjoyed special access and privileges normally reserved for governments alone, including sharing building facilities with UNESCO in Paris. Second, in 1968, UNESCO, along with several other specialized agencies convened the Man and the Biosphere program that focused on conserving natural resources on an ecosystem level.[2] The program promoted the protection of ecosystems through the biosphere reserve system, administrated through the appropriate national jurisdiction. Registration and administration in the biosphere reserve system has remained voluntary and at the sole discretion of the individual nation. Unlike the latter ad hoc conference system, this convention had no ambitions towards negotiating formal treaties or forming "common property" regimes. Instead, the program focused on voluntary registration and exchange of information and ideas, with an eye towards establishing best practices for resource use and conservation. However, these isolated incidents never fully captured the political imagination or resources in order to secure environmental issues a permanent, prominent place on the UN agenda, although they did serve to begin the march towards global governance with its emphasis on ideas and non-state actors.

Sustainable development, however, did become the cornerstone of global environmental politics. Although academic research typically gives the history of sustainable development as beginning with the Brundtland Commission by way of an introduction to sustainability, (Bartelmus 2003; Goodland 1995; Howarth 1997; Lafferty and Meadowcroft 2000; Lipschutz 2004; Middleton and O'Keefe 2003; Rao 2000), the fundamental building blocks of this ideology began much earlier (Irwin 2001; Selin and Linnér 2003). The Brundtland Commission did not merely occur at a particular time and place in history, but rather is itself a function of political necessity. This story, then, begins by asking why the UN system assembled the Brundtland Commission on sustainable development. The answer to this question has its roots in the events surrounding the first of the major UN conference – the UNCHE.

---

[1]For more information on the formative years of UNESCO, see Hoggart (1996). Additionally, Boardman (1981) provides a historical narrative of the first 30 years of IUCN.

[2]The other sponsoring agencies included the Food and Agriculture Organization, the World Health Organization, the IUCN, and the International Biological Program.

4.2 Stockholm

Environmental affairs came to the forefront of the UN system in 1968 when the Swedish government introduced a resolution calling for a conference on the environment to "dramatize" environmental degradation that was deemed to require international cooperation and action to mitigate further damage (Engfeldt 1973:394).[3] The resolution suggested that it was "desirable to provide a framework for comprehensive consideration within the UN of the problems of human environment in order to focus the attention of Governments and public opinion on the importance and urgency of this question" (UN 1968b). The next year, the UN General Assembly finalized a structure for the conference and also asked the Secretary-General of the Conference, Maurice Strong, to "take into account the results of other international conferences on the subject and to draw on contributions from appropriate organizations [and] invited the specialized agencies and other organizations concerned to assist in the work of the Preparatory Committee" (UN 1969). The Resolution further stipulated that the conference was to last for 2 weeks and accepted Sweden's invitation to host the event.

However, the road to Stockholm was neither smooth nor straightforward. What began as concern about the deteriorating physical characteristics of the environment, as articulated by the North, gave way to the South's deep desire for ensuring future development. The process of incorporating a Southern view into the environmental conference emerged when developing countries, along with the more socialist-leaning developed nations, were initially unwilling to participate in the conference since it did not deal directly with this theme.[4] Southern discontent with the emphasis on environmental degradation was compounded by the addition of Communist China to the G-77 in 1971 and a bid by Brazil to assert its particular preferences on the other members of this group. Brazil strongly opposed the conference as a concern of the wealthy nations. If pollution was the result of industrialization, then Brazil wished for pollution. Accordingly, industrial development was the first priority for the G-77 and China.

Nowhere was this demand for greater attention to the Southern agenda more militant than in General Assembly Resolution 2849 (XXVI) of 20 December 1971. This resolution injected the G-77 position into the Stockholm Preparatory Committee process and blamed environmental degradation on the developed countries. Consequently, the resolution's authors called on these countries not only to clean up their own internal environment, but to pay for the environmental damage within the developing countries as well. Further, the resolution stated that environmental protection activities must not impose further suffering on the developing nations,

---

[3]Sweden was on the receiving end of air pollution from some of its neighbors, particularly the United Kingdom, that was perceived as killing its forests.

[4]Landsberg (1972:749) bemoans the increasing demands by the South, especially an attempt by some African countries to use Stockholm as an opportunity to demand reparations for apartheid and natural resource use.

most especially through the use of environmental standards that would block the export of goods from the developing countries to the developed countries.[5] As the developed countries were not inclined to accept this position, much less provide the significant amounts of funding that environmental restoration would have required, this issue had the potential to derail the conference before it began.

Maurice Strong of Canada served as Secretary-General for the conference. Strong was not the original choice to lead the Conference as the position first went to Sweden's Jean Mussard. He proved to be unable to provide the strong personal leadership necessary to head off the pitfalls surrounding the conference. This deficiency was perhaps most apparent when he did not require the other specialized UN agencies to funnel their background reports and other essential information through the conference secretariat, effectively losing control of the political process; nor did he attempt to convince the developing countries to participate fully in the conference. Strong, in contrast, embodied a strong personal leadership style and a background in developmental aid. The latter, in particular, proved farsighted, as Strong's familiarity and competency in developmental politics allowed an issue linkage between environment and development that enabled the conference arrangements to proceed.

Strong (2000) reports in his autobiography that he linked the environment and development issue areas together in order to obtain the political support necessary to convene the Stockholm Conference. The thesis sometimes referred to as ecodevelopment conceptualized environmental quality as a basic necessity for future development.[6] Environmental degradation was not a "quality of life" indicator for the South, but rather a threat to life itself. In other words, underdevelopment and lack of local control of natural resources caused environmental degradation in the South, while industrialization caused environmental degradation in the North. Thus, ecodevelopment went on to claim that further industrialization would be the best cure for environmental problems in the Third World, while implying that decreasing industrialization and redistributing natural resource patterns would solve the Northern pollution problem, although consumption and production patterns were never explicitly discussed as part of the Stockholm conference.[7]

The Panel of Experts on Development and Environment, held at Founex, Switzerland from 4 to 12 June 1971, further refined Strong's ecodevelopment thesis. The Panel met to discuss environmental considerations in the context of economic development, per Southern demands (Caldwell 1984). These countries viewed environmental degradation as a topic for the wealthy nations, at best, and at worst, a perceived threat to impose a new neo-colonialism upon the South

[5] See Luchins (1977) for a detailed history of the Stockholm Conference.

[6] See Strong (1972) for clarification of the ecodevelopment concept and for Strong's expectations for the Stockholm Conference as a whole.

[7] As the Stockholm conference took place in the middle of the Cold War, albeit in a period of détente, it should not be surprising that issues of political economy would be kept off of the negotiating agenda as this would have immediately ended the possibility of consensus.

(Birnie 1993; Castro 1972; Holdgate et al. 1982:6; Johnson 1972; Kennet 1972:37; Landsberg 1972:749).[8] The ecodevelopment thesis garnered important support from developmental economists Barbara Ward (United States), Gamani Corea (Sri Lanka) and Mahbub ul Haq (Pakistan) who worked together in the months preceding the Founex meeting to flesh out the details of this position.[9]

The Founex report distinguished between the environmental problems of the industrialized North and the underdeveloped South. It claimed that production and overconsumption of goods caused Northern environmental problems, while poverty and underdevelopment were at the root of environmental problems in the South. Correspondingly, "There is a pollution of affluence and a pollution of poverty" (Castro 1972:409).[10] Northern consumption patterns and unequal wealth distribution bore responsibility for environmental pollution in both the North and the South. Ambassador Castro's statement, particularly the notion of "the pollution of poverty," became a rallying cry for the Southern countries that clearly stated a desire not to be construed as Rousseau's "happy savage."

As the Report of the Panel of Experts on Environment and Development (UN 1972a) states in its opening chapter:

> [T]he major environmental problems of developing countries are essentially of a different kind. They are predominantly problems that reflect the poverty and very lack of development of their societies. They are problems, in other words, of both rural and urban poverty. In both the towns and in the countryside, not merely the "quality of life," but life itself is endangered by poor water, housing, sanitation and nutrition, by sickness and disease and by natural disasters. These are problems, no less than those of industrial pollution, that clamor for attention in the context of the concern with human environment. They are problems which affect the greater mass of mankind.

While this statement, and indeed the whole report, does not refer to ecodevelopment as a distinctive ideology, concern with the future tenets of sustainable development – environmental protection, underdevelopment, and social justice are present, albeit not in a form that would be associated with sustainable development despite later UNEP claims. Instead of treating the Founex report as an enlightened precursor to sustainable development, it is more correct to think of it as the developing countries' deep-seated fear and concern that environmental politics would restrict the South's desire for industrialization. Thus, the Founex report is a limited perception of environmental affairs instead of a widely accepted ideology. It deals specifically with the need to ensure that environmental protection does not hamper industrial development and, in cases where environmental protectionism halts or makes industrialization more costly, it seeks monetary compensation for the loss

[8]Gardner (1972:240) disagrees, citing Brazil as the reticent state, "which purports to see in international discussion on the environment a sinister plot against its own development."

[9]For a list of conference participants, see UN (1972a).

[10]Joao Augusto de Araujo Castro served as the Brazilian Ambassador to the United States during the time of the Stockholm Conference. While this statement cannot be termed an "official" position of Brazil, it is nevertheless strongly reflective of the Southern hemisphere's thoughts on the Conference.

of development potential. Further, the document's impact was limited because development represented a small portion of the overall agenda.

Despite Strong's careful work on the ecodevelopment thesis, not all of the developing countries endorsed his efforts wholeheartedly. India, Algeria, and Bangladesh took time to question the nature of environmental degradation and its links to development. Ecodevelopment attempted to place environment and development on equal footing, contrary to the wishes of many of the developing countries, which preferred to focus solely on development. In a speech that clearly displayed the dominance of developmental affairs over environmental protection, the Algerian government questioned:

> But what would be the use of restoring nature in a world where man remained oppressed? What would be the use of conserving natural recourses in a world dominated by economic inequality and social injustice? What could be the use of a newly viable environment if the majority of human societies continued to have no say in the major decisions that govern the world and to be subject to arrangements and compromises concluded over their heads? (Algeria 1972: 114)

The Algerian government was not concerned with environment per se, but rather with the broader socio-economic concerns between different "classes." Despite the South's lukewarm support of ecodevelopment, the Founex Conference served its purpose in attracting the attention of the South, allowing the conference to proceed.

Sweden hosted the UNCHE in Stockholm on 5–16 July 1972. Officials from 113 states converged on the city for this occasion, including the Prime Minister of India, Indira Gandhi, and the host of the conference, Olof Palme, Prime Minister of Sweden.[11] Palme (1972:52) defined the work of the conference in his welcoming speech by proclaiming:

> But if this conference can outline a clear course of action and if it can create organizational procedures for further work on environmental problems, then I am sure that it will be regarded as the beginning of a new stage in international cooperation. It will have created confidence in the future.

History judged Stockholm a success in this regard as it produced three noteworthy features of international environmental politics, in addition to the NGO-inspired moratorium on whaling. More generally, Stockholm was the last conference that East-West politics threatened to split and the first to foreshadow the North-South gap that would take its place in UN politics. Ambassadors and scholars alike recognized the uniqueness of the Stockholm Conference when it occurred, for what Stockholm sought was nothing short of building a new infrastructure on which to construct the norm of environmental protection.

The first and most significant action was to recommend the creation of UNEP. The General Assembly acted on this recommendation in Resolution 2997 (XXVII)

---

[11] The Communist states, led by the Soviet Union, boycotted the Conference because West Germany attended, while East Germany was not present. Nevertheless, Soviet officials coincidentally visited Stockholm during the Conference and Secretary-General Strong provided discreet shuttle diplomacy services to ensure that Conference documents reflected the Soviet block's views.

in 1972. UN (1972b) was expected "to safeguard and enhance the environment for the benefit of present and future generations of man," by coordinating environmental policies and activities taking place in other UN bodies. Structurally, it consisted of a 54-member Governing Council that reported to the UN's ECOSOC; an Environmental Secretariat, headquartered in Nairobi, Kenya; an Environmental Coordination Board; and the Environment Fund, a voluntary fund used to finance environmental initiatives both within and beyond the UN system. Second, UNCHE produced an "action plan" for national governments to consider when establishing their own environmental ministries. UNEP documents report that governmental bodies with responsibility for the environment grew from 10 to 100 in the first 10 years after Stockholm, with a corresponding increase in legislation passed and in NGOs' activity related to environmental management. Third, UNCHE called for the creation of a global monitoring system, now known as Earthwatch. Earthwatch provides an integrated system of data gathering and dissemination and serves as the primary focal point for the dissemination of scientific and socio-cultural data on environmental affairs, including sustainable development.[12] Earthwatch does not refer to a single database, but instead to a collection of data made available through the UN system. Monitoring projects include background air pollution, ozone depletion, hazardous waste, climate change, oceans, and coastal areas. Earthwatch also hosts the International Register for Potentially Toxic Chemicals. Equally noteworthy, however, is the fact that Stockholm did not address the underlying socio-economic structural problems that caused environmental degradation such as inappropriate land use, natural inequitable resource distribution, and alternatives to large capital infrastructure projects as a means to achieving industrialization (Feraru 1985:56). The failure of these items to appear on the conference agenda, however, does not negate the fact that the conference generally exceeded expectations in the results it produced.

The dominance of development, particularly industrialization, over environmental protection was the majority opinion among the developing countries. Of the 109 Stockholm recommendations stated in the Stockholm Plan of Action, two carry particular weight in illustrating this point—compensation, in Recommendation 103, and additionality, in Recommendation 107. The Stockholm Action Plan articulated the compensation recommendation as, "That where environmental concerns lead to restrictions on trade, or to stricter environmental standards with negative effects on exports, particularly from developing countries, appropriate measures for compensation should be worked out within the framework of existing contractual and institutional arrangements (UN 1972c)." In other words, developing countries had a right to legal recompense for the decrease in export earnings caused by increasing environmental protectionism by the importing country. Additionality articulated

[12] I am referring here to the UN systemwide Earthwatch and not to the NGO Earthwatch Institute, which uses the same moniker. Earthwatch under UNEP auspices also should not be confused with the Worldwatch Institute established by Lester Brown during the same general time frame, although there exists important collaborative ties between the two entities.

the principle that environmental regulations should not impede financial flows to developing countries and that new assistance was required to enable less affluent nations to protect their environment. In other words, developing countries needed more money to implement new environmental regulations. These two elements combined in an early attempt by the developing countries to create an international obligation for financial flows from the North to the South. Two General Assembly resolutions passed in 1974 would make this viewpoint clear – the New International Economic Order (NIEO) and a Charter of Economic Rights and the Duties of States. NIEO clearly stated that state sovereignty included the right to use natural resources at the discretion of the territorial state, as long as such action did not damage the legitimate interests of other states.

The legacy of ecodevelopment stems from the linkage between the North's technocratic conceptualization of environment and the South's preoccupation with funds for industrialization. The North acquiesced to this tradeoff when they acknowledged the legitimacy of providing financial aid to the South to catalyze development in an environmentally friendly manner in exchange for the acquiescence on the part of developing countries for increased measures of environmental protection. In time, the compromise that member states negotiated for Stockholm would grow into a new rhetoric within international politics, although this process would not occur quickly, smoothly, or uniformly.

Ecodevelopment was not the only new aspect of international environmental governance brought to light at Stockholm. NGOs also contributed actively to the Stockholm conference, although there is considerable discrepancy about the size and actions of the NGO community.[13] Strong and his staff met prior to the Stockholm conference with important NGOs, such as the ICC, to solicit their advice and to encourage them to participate in the political process both before and during the Stockholm Conference (Herter and Binder 1993:26). It seems likely that Strong was engaged with a broad-based spectrum of NGOs, including the ENGOs such as Greenpeace and Friends of the Earth. Ward and Strong met in Canada, which led to both a professional working relationship and an enduring friendship that ultimately contributed to the progressive elucidation of sustainable development.[14] Indeed, it was Strong who made the initial appointments that led to Ward overseeing the rehabilitation of the International Institute of Environmental Affairs (IIEA) into the highly influential International Institute for Environment and Development (IIED).

Secretary-General Strong set yet another UN-NGO interaction milestone when he requested that Ward, along with Dr. René Dubos chair a group of experts for

[13]The growing awareness of environmental affairs in the North, which in turn engendered support for the NGOs was spurred on by important books such as *Silent Spring* (Carson 1962), *The Limits to Growth* (Meadows et al. 1972), *The Closing Circle* (Commoner 1971), *This Endangered Planet* (Falk 1971), and *Only One Earth* (Ward and Dubos 1972).

[14]Despite the fact that Ward served as Strong's senior adviser at the Stockholm Conference, both the Strong and Ward archives are curiously devoid of personal writings from this 2-year period and only a handful of letters exist between the two in 1977. The letters are located in the Barbara Ward archives at Georgetown University, Box 3, Folder 19.

the purpose of writing a background report for the conference. An NGO, the International Institute for Environmental Affairs (IIEA) oversaw the preparation of this report, better known by its published title – *Only One Earth*.

In fact, NGOs provided a necessary labor force for the extensive preparatory work. IIEA "marshaled the agenda for the Stockholm Conference on the Human Environment (Cross 2003:xiii)." IUCN and the International Council for Scientific Unions (ICSU, now International Council for Science), in particular, played important roles in producing reports for the Stockholm conference (Bazell 1972:390; Gardner 1972:241). ICSU formed a subcommittee, Special Committee on Problems of the Environment (SCOPE), that acted as a consultant for the conference secretariat. SCOPE produced reports on pollutant monitoring, international registry for chemical compounds, and implications of "man-made" ecosystems (Bazell 1972:391). IUCN reported that it was "advising" on almost every agenda item for the conference, despite the fact that some of this advice appeared to be unsolicited. At the time of the UNCHE the role of NGOs acting as consultants to the conference secretariat was a radical departure from customary UN-NGO relations. Previously, NGOs had not played substantive roles in UN conference negotiations, in spite of an oft-expressed desire of the NGOs to participate.[15] NGOs were able to "break through" the sovereign state barrier because of their unique technical experience with the environment, as well as the strength of the political pressure the ENGOs were able to bring upon the conference, as the IUCN was not alone in giving advice to the conference secretariat. This political pressure took multiple forms including the provision of consulting services, raising the profile of the conference through letters to the editor and other public reports, hosting events in support of Stockholm, and publicizing NGO expectations as to the outcomes.

Moreover, NGOs did not end their novel role at the beginning of the formal conference in Stockholm. Various reports suggest that between 255 and 298 NGOs attended the official NGO meeting (Feraru 1974; Willetts 1989). Additionally, several smaller but more colorful unofficial conferences also occurred in Stockholm around the same time as the UNCHE.[16] According to Stone (1973:17),

> Stockholm had many of the attributes of a theatrical festival, and a very modern one at that. There were many departures from the scripts, a number of happenings, the audience participated now and then and there was even a bit of off-stage nudity. When it was over the critics reviewed it in the Press, praising the sets and the production, and commenting on the interpretations by the various heroes, villains and bit players. Dramatic tension was maintained to the very last hour of the last day.

---

[15]The World Federation of Trade Unions precipitated an early crisis within the UN in 1946 when it asked for speaking privileges essentially equal to states. Several other NGOs immediately followed suit, including the ICC. The UN member states denied all of the petitions, but NGOs won limited speaking opportunities.

[16]Stone (1973:62), the Senior Information Adviser to the Secretary-General would later write "unfortunately in the environmental field the strength and value of NGOs tends to be inversely proportional to their acceptability in the international scene."

The Preparatory Committee for Stockholm, and its Secretary-General Maurice Strong, made arrangements for NGOs to attend a parallel conference dubbed the Environment Forum. Three other parallel conferences sprang up—one organized by Dai Dong involving a group of 30 scientists funded by Xerox; one called the Folkets Forum, an event organized by a variety of Scandinavian groups; and one the Life Forum, which was better known as the Hog Farm that mimicked American counter-culture groups in large part because it was organized by Stuart Brand, a Merry Prankster and founder of the Whole Earth Catalogue (Rowland 1973).[17]

The Environment Forum was the largest of the NGO conferences occurring simultaneously with the Stockholm meetings. It was born of political necessity to circumvent criticism of the conference by NGOs. The conference secretariat also expected the sizeable number of NGOs to promote the conference if they were invited to attend. These participants were accredited as observers to the formal meetings and had access to governmental delegations and conference staff through close proximity, as well as rare speaking opportunities. The Environment Forum was not entirely successful in its attempt to entice NGOs into the UN sphere of influence as evidenced by the three other major conferences and multiple individual events that were held during this time.

The Environment Forum quickly split into two groups. The parallel conference was run by the National Council of Swedish Youth and the Swedish UN Association. This forum included a multilingual reference library, a film program, and exhibit facilities (Berg 1972:746). A poor choice of coordinators set the stage for a de facto coup by Barry Commoner and his associates. The Forum quickly descended into a chaotic debate that included aspects of Commoner's personal agenda such as protests against United States involvement in Vietnam and the promotion of an idiosyncratic version of "semi-Marxism" (Eco 1972b:59). Berg, a colleague of Commoner's who also attended the conference, denied that Commoner was orchestrating the Forum. Instead, Commoner attempted to refrain from engaging in political debates (Berg 1972). Regardless of Commoner's intent, freedom of speech at the Forum was questionable and violent outbreaks occurred, most notably during a conference on population headlined by biologist Paul Ehrlich (Artin 1973; Stone 1973). The official conference quickly turned into a safe haven from the undiplomatic behavior occurring at the Forum. News items from the Forum were dutifully reported in the Stockholm Conference *Eco* and in the other major daily newspapers.

The second, and generally more respectable, Dai Dong conference brought together 29 scientists from 24 nations to negotiate an environmental treaty. [18] Dai

[17] Artin (1973) writes primarily about the Dai Dong International Fellowship of Reconcillation conference, although he also provides anecdotes about the Hog Farm, the Environment Forum, and the formal conference as well.

[18] Dai Dong had loose connections to the International Fellowship for Reconciliation (Artin 1973). It was a short-lived project (1969–1975) that attempted to promote both peace and the environment. The Dai Dong conference at Stockholm proved to be the most noteworthy event in the history of this organization.

Dong expected developed countries and industrial concerns to dominate the official conference. They sought to highlight the positions of underdeveloped countries by negotiating an alternative treaty text. The Dai Dong conference occurred in the 2 weeks before the Stockholm conference. The conference originally intended to have a draft declaration and a series of concrete policy recommendations. Despite the Dai Dong secretariat's preparation of a draft declaration, the conference participants saw fit to reopen the declaration for discussion. As a result, the conference found itself mired in controversy along the lines of the North-South gap, although Artin (1973) did not recognize this pattern despite his documentation of it. Three main items of contention included the denunciation of the principle of national sovereignty, condemnation of all wars, (including wars of liberation) and population control.[19] Additionally, Dai Dong's treaty suffered from polarized political rhetoric with six of the scientists agreeing to sign the document only after adding a statement repudiating large portions of the text. Consequently, Dai Dong barely finished negotiating the draft declaration before its release at a prescheduled press conference and the conferees failed to discuss any of their policy recommendations as had been originally intended.

Not surprisingly, the Dai Dong treaty received mixed reviews at best. The Stockholm Conference Eco (1972a:14) authors praised the document and noted that the treaty is supported by those who have low expectations for the conference outcomes. *Eco*, however, did not endorse any specific recommendations within the treaty. Rowland (1973:128) proclaimed the treaty a disaster by pointing out that not only did the document fail to reach agreement on important topics such as population, but the "apolitical" scientists gathered by Dai Dong failed to achieve the same level of agreement as the official conference.

The Folkets Forum, or People's Forum, was sponsored by several Scandinavian groups. Pow Wow, the steering committee that organized this event, was a loose collaboration of left-wing environmental and political groups. As such, it covered a large range of topics including ecocide, drugs, and Chinese environmental policy. The forum did not suffer from the organizational chaos experienced by its colleagues. Its admittedly Marxist philosophies attacked the UN conference for failing to deal with the underlying problems of environmental degradation – limits to growth, lack of population control, and unequal distribution of resources (Stone 1973:131).

The Life Forum group was the most colorful of the parallel conferences. Participants met at an abandoned Stockholm area airport (*Skarpnänck*) that the media dubbed the Hog Farm, after the American Merry Pranksters that organized the event, which dominated headlines from the Life Forum.[20] The Swedish government

[19]Dai Dong was not able to reconcile divergent views on population. The final draft of its declaration stated its belief in the limit of population growth, but the statement makes no reference to population control.

[20]The Merry Pranksters, led by Ken Kesey, epitomized the radical counter-cultural movement of the 1960s. Kesey and his associates led the American cultural experimentation with LSD and

lent the land to the NGOs for the duration of the conference. The government feared a Woodstock-like festival and hoped that the Merry Pranksters would be able to police the area. The atmosphere at the Hog Farm at times represented more of a party, with its "street theatre, rock concerts and dress-up parades" (Rowland 1973:1).

While the "entertaining encouragements" were "unnecessary" according to Rowland, Strong (2000:132) disagrees. NGOs effectively lobbied for a moratorium on whaling, by raising the status of this issue through the world media that had also converged on Stockholm. He further reports that NGOs ability to impact public opinion helped to raise environmental consciousness, particularly in developing countries. It also allowed them access to decision makers as the public expected NGOs to be part of the international negotiating processes. According to this interpretation, Stockholm was the first time ENGOs could successfully integrate their agenda into an international conference. The continuation of the practices begun in Stockholm with respect to the NGOs would, in time, revamp the international system as it existed in 1972.

Stockholm, then, can be interpreted as establishing the framework for environmental affairs for the next 30 years. On one hand, the developed North defined environment in biophysical terms and environmental degradation as the increasingly negative changes to the Earth. The underdeveloped South, on the other hand, alternated between the potential for new financial aid to "green" their development strategies and the fear that the new environmental standards would lead to increased economic protectionism and less demand for their goods. They saw the North as the primary agent of environmental damage, either directly by overconsumption, or indirectly through the perpetuation of the socio-economic status quo. Accordingly, the South had more pressing social concerns than environmental protection, namely development, which in turn added to environmental pollution either through the increased presence of heavy industry or through the conversion of natural resources into commodities, overwhelmingly for consumption in the North.

The compromise that occurred at Stockholm sought to promote the Northern version of environmental protection, while at the same time allaying Southern fears about negative economic impacts. While ecodevelopment served its purpose in bringing the G-77 to the negotiating table at Stockholm, delegates enacted few concrete action items to support ecodevelopment within the Stockholm Action Plan.

The detailed discussions and immense scope of work for this conference created a greater role for NGOs, giving this agent direct access to promote ideas into the formal structure of the "state system". Efforts to explore, then reconcile, the North-South gap leading up to the Conference gave NGOs the opportunity to shape the conference through their advice, both solicited and unsolicited. Spiro (1994:50), for example, suggests that IUCN provided an early draft for the Stockholm Declaration

marijuana, amongst other psychedelic drugs. While the Merry Pranksters flaunted their use of dangerous and illegal drugs, their followers were generally well behaved.

and this allowed the organization to influence directly the proceedings of the event, if not the direct outcome of the Stockholm conference.

The role and function of NGOs changed dramatically because of the Stockholm Conference. Prior to this time, NGOs thought of themselves as predominantly "apolitical" organizations whose purpose was to advise UN delegates (i.e., states). Stockholm brought a new, more activist group of NGOs into UN politics. After Stockholm, NGOs as a collective would no longer be willing to serve the narrowly constructed apolitical role established for them by UN diplomats.

ENGOs exemplified this shift in activity. Strong believed in a political process that included scientific expertise and, to tap into this scientific expertise, he created multiple international working groups consisting of both public and private professionals. Strong requested and received national environmental reports from 85 countries, some of which did not have an environmental ministry. Thus, his request had the effect of raising the political relevancy of environmentalists and scientists within their home countries as many of these environmentalists and scientists interacted with their state governments for the first time. Additionally, it seems likely that the shortened time frame prompted by Mussard's replacement with Strong also played into the necessity of involving NGOs' specialized knowledge because it halved the period for national governments to produce their national reports from 3 years to 18 months.

Once Strong had mobilized this scientific expertise, it is likely that these organizations and individuals gave more credence and importance to the conference, in turn, making it more likely that they would participate at the UN in the future. For example, these groups continued to lobby national governments after the Stockholm conference in support of the Stockholm outcomes – to support the establishment of UNEP, to carry out the Stockholm Action Plan, to stop offshore drilling, and to stop killing whales.

ENGOs were not the only groups active at the Stockholm Conference. Their business counterparts, because of their inherent conservatism, were, however, more likely to be found outside the conference rooms than in separate forums. That is not to say that BINGOs did not host their own affairs. Artin (1973:7) complained that one of Sweden's motivations in hosting the conference was to stage a public relations coup that included promoting Sweden's pollution-control technology industry, spearheaded by the Federation of Swedish Industries. The more entrepreneurial companies took out advertising in the Stockholm Conference *Eco* to proclaim the benefits of their new technologies. Additionally, firms lobbied delegates by hosting special events, including plant tours.[21]

At the time of the conference, it was not clear that NGOs had any role to play in formal negotiations, despite the fact that they were allowed to directly address the main committee. Eco (1972d:53) ran an article entitled "NGOs Go Home"

[21]Companies with advertisements for special events in *Eco* are Philips (3), Atlas Copco (2), Pollution Technical Services Limited (3), Wilderness Expeditions (1) and Englehard (1). Additionally, Stone (1973: 137) reports that Volvo hosted an event at its facilities in Gothenburg.

bemoaning NGOs' lack of positive contributions to the UN negotiating session. [22] The next day the headline read "NGO come back, then go home" (Eco 1972c). The article explains that conference participants did not appreciate NGOs commitment to the environment or the level of their sophistication and understanding of the diplomatic processes. While the specifics of this speech have been lost, Margaret Mead apparently spoke on behalf of a wide-ranging coalition of NGOs, and was likely concerned with both substantive and procedural issues.[23] She went on to chair a meeting of NGOs that pledged to work together in a loose confederation to interface more effectively with the UN system. Thus, one of the lessons learned from Stockholm is that NGOs that are able to coordinate their policy recommendations and lobbying activities are more likely to influence governments at the UN than their more colorful counterparts.

However, the more animated NGOs were probably more effective at attracting media support as the Moratorium on Whaling suggests. In this case, the NGOs loosely associated with the Hog Farm created a car in the shape of a whale and paraded it in and around Stockholm during the conference. This theatrical stunt focused the attention on whaling, an issue not originally on the conference agenda. The United States introduced the resolution, one that the conference passed despite significant opposition from Norway and Japan. NGO activists swayed public opinion in favor of this initiative through a variety of measures. First, they unilaterally spoke in favor of the ban. Second, they adopted the whale as a highly visible symbol of the conference and this gesture, in effect, turned the success or failure to agree to a whaling moratorium into a litmus test for the success or failure of the conference as a whole. The Stockholm Conference conveyed its political support for the 10-year moratorium on commercial whaling to the International Whaling Commission (IWC) though the IWC ultimately declined to implement the ban, citing a lack of scientific evidence.[24]

The above examples of the expansions of NGOs into UN politics through the provision of scientific expertise, the mobilization of both media attention and public support, and the incorporation of activist organizations as a "constituency" of the UN indicate a clear expansion of role for the previously lackluster "apolitical" entities that participated in formal international affairs. Nevertheless, it was not clear at the time that these changes would be permanent, in large part because much of the activity occurred away from the formal negotiating roles occupied by member

---

[22]ECO was published by T*he Ecologist* (funded by Teddy Goldsmith) as part of environmental NGOs political lobbying activities. The paper generally supported pro-environmental positions, causes and delegations.

[23]Artin (1973:163) provides a partial list of NGOs affiliated with the speech including the Sierra Club, International Planned Parenthood Federation, Congress of African People, World Wildlife Fund, Socialist International, European Oceanic Association, ICC, International Alliance of Women and United World Federalists.

[24]Japan abstained from the vote on this issue.

states. That is, the expansion of NGOs scope occurred "off the record" so to speak, and this precedent might not have carried over beyond this conference, much less into other UN political arenas.

The structures of global environmental politics also changed as a result of the Stockholm conference. Most notably, Stockholm marked the expansion of the state system through the creation of UNEP. Equally significantly, the politically active NGOs also used Stockholm to network with each other, including the formation of an NGO coordinating center, originally known as INASEN, standing for the International Assembly of Non-Governmental Organizations Concerned with the Environment. While INASEN was short-lived due to difficulty legally registering with the Kenyan government, it transformed itself into the more-widely known ELCI within 3 years of its establishment. INASEN formally dissolved in 1977, in large part because the large ENGOs preferred to interface directly with UNEP and had no need for INASEN to act as an interlocutor.

4.3 The Lost Years

Stockholm's resolutions were forwarded to the General Assembly for approval at the end of 1972. The Conference report was endorsed in Resolution 2997 (XXVII) and UNEP was established with Maurice Strong as its first executive director. While Strong agreed to serve in this capacity, he announced that he would only stay for a year. In an unexpected move, the resolution also named Nairobi, Kenya as UNEP's headquarters.[25] The selection of Nairobi as the host city heralded the newfound political strength of the G-77 block. Further, the General Assembly assigned UNEP a catalytic and coordinating role, much weaker than many at Stockholm had envisioned; weakened, no doubt by the Brussels group, a collection of industrialized countries that had worked behind the scenes to ensure that UNEP would not be able to utilize environmental protectionism to harm their economic priorities.[26] While not documented, it seems likely that the G-77's preference for Nairobi likely found support, or at least limited resistance, from the Brussels group. UNEP's primary function has been to coordinate activities implemented by other UN bodies, an oft-cited weakness of the international governance system. As a coordinating body, it does not automatically take the lead during treaty negotiations, although it has served a prominent role in this respect in the Regional Seas programs and the Montreal Protocol to protect the ozone layer. This catalytic function arose in part out of governmental concerns that a new agency would disrupt work by other,

[25] At the conclusion of the Stockholm Conference, London and Geneva were early favorites for the privilege of hosting UNEP. Nairobi emerged as a candidate city during the 1972 General Assembly session.

[26] The Brussels group included Britain, France, the United States, West Germany, Italy, Belgium and the Netherlands. See Hamer (2002) for further explanation.

more politically astute, agencies already engaged in environmental policies such as the World Health Organization and the Food and Agriculture Organization (French 1995:28). Governments intended for UNEP to coordinate information gathering and policy development, and to serve as a source of (limited) funding for projects through the environment fund.[27] As UNEP is a political body, all reports they issued had to be politically feasible and this excluded a great deal of new information and research necessary to further the Southern environmental agenda. In order to solve this dilemma, Strong, in 1974, urged the IIED to comment upon his strategy of funding independent organizations to provide information that would serve to raise public awareness (Tinker 2003:40). IIED, not surprisingly, enthusiastically endorsed the idea and immediately requested funding for a trial period, which UNEP provided.

The "capture" of UNEP by a developing country served notice that the G-77 intended to impose on the new organization its own particular vision of environmental affairs. The developing countries wanted to ensure that environmental policy did not block further development. Rather, increased industrialization was necessary to reduce the "poverty of pollution." This objective stood in contrast to the Northern preference for studying and conserving the physical characteristics of the environment. At the very least, the removal of UNEP from the traditional corridors of UN political-decision making lessened the chances that environmental protection would interfere with continuing Southern industrialization. To a certain extent, the G-77's last minute proposal to locate UNEP in Nairobi served as a striking symbol of various government commitments, or lack thereof, to the environment. UNEP's physical proximity to the South ensured that environmental protection would not be used as an excuse to halt developmental aid, while at the same time insulating Northern countries from their responsibility for the global environmental crisis. UNEP's location in Nairobi, however, had an unintended side-effect – of drawing Southern concerns (i.e., development) into a tighter relationship with the environment. The need for Southern solidarity to protect the only UN headquarters outside of the developed countries overrode individual state preferences (Gosovic 1992:11). Consequently, the South's great fear, that UNEP would advance environmental protection at the expense of Southern development, instead created an ideal situation where a new institution was located on the fringe of the UN machinery, with a relatively weak mandate. Over time, this situation led to a close working relationship between UNEP and the developing countries as well as resulting in UNEP reliance upon NGO resources.

Under these circumstances, it is perhaps not surprising that although UNEP played a coordinating role within the UN system and helped to garner attention for environmental issues, the organization gained a reputation as being representative of an eclectic constituency led by developing countries and NGOs. After all, UNEP

---

[27] Within the UN system, the UN Development Program, along with the Bretton Woods organizations, serve as the primary conduit of funds to the developing countries.

was headquartered in a developing country (Nairobi), led by Mustafa Tolba, a member of a developing country (Egypt), and had a governing council that enforced a developing country majority.[28]

It was thus immediately after Stockholm that the North-South gap that began within the preparatory conferences meetings began to harden. After Stockholm, international environmental agreements continued to multiply. Technical accords such as the Convention on the International Trade in Endangered Species and the Convention on the Prevention of Marine Pollution by Dumping of Wastes and Other Matter were rapidly negotiated and implemented. However, more subjective issues pertaining to housing and human settlements, water, and desertification were unable to attain independent regime status.

Ecodevelopment, however, made a major step forward in its evolution towards sustainable development with the advent of the Symposium on Resource Use, Environment, and Development Strategies stemming from a meeting jointly sponsored by the UN Council on Trade and Development (UNCTAD) and UNEP in Cocoyoc, Mexico, in October, 1974 (Adams 1990:40). Unlike the Founex conference preceding it, Cocoyoc provided the first look at the environmental issue arena from the perspective of the impoverished populations of the Third World. The final document, drafted by Barbara Ward, criticized the overconsumption of the North in comparison to the absolute poverty of the South. The document introduced such phrases as the "inner limits" of human needs, i.e., food, shelter, clothing, education, and health, and the "outer limits" of natural resource constraints. Cocoyoc placed the eradication of poverty at the center of the intersection between environment and development through its insistence on multiple pathways to development based on local culture and resources. Not surprisingly, Cocoyoc, in many ways, highlighted the key points of the NIEO rhetoric popular at this particular point of UN history. Cocoyoc, like the NIEO, never made it past the phase of interesting, if naïve, oratory. As Tinker (1975:481) reports, "American diplomats have played lip service to views on development which they emphatically do not share." Nor were the Americans alone in their lip service to the NIEO, as the overwhelming majority of the developed North shared in the sentiment, although were perhaps not as vocal in their opposition.

Cocoyoc's rhetoric may have been politically unfeasible, but it did add a key concept to ecodevelopment – the need for social justice including access to food, clothing, shelter, health care, and education. It thus moved ecodevelopment a step further away from focusing solely on the economic distribution of wealth towards a broader focus on current human needs.

While Cocoyoc proceeded to highlight the radical rhetoric associated with both ecodevelopment and the NIEO, the yearly sessions of the UNEP Governing Council moved forward in a much calmer fashion. Likewise, NGOs interested in the

---

[28] Of the original 58 member seats on the Governing Council 16 of them were for African countries, 13 for Asian countries, 10 for Latin American countries, 13 for Western Europe and other states and 6 for Eastern Europe.

environment continued their close contact with UNEP. Larger organizations did not need a coordinating body to interface with officials in Nairobi as they were easily able to act on their own behalf, despite the geographic distance.[29] Nevertheless, the capability to remain in close contact with UNEP was aided greatly by ELCI as it shared an office with UNESCO-Africa in Nairobi, and this location gave it excellent geographic proximity to the UNEP headquarters. Environment Liaison Board (ELB), in particular, enjoyed a close working relationship with UNEP that included speaking at the UNEP Governing Council Sessions. Text from the 1976 ELB speech delivered by Cyril Ritchie, a member of the Board of Directors of ELB, introduced the concept of intergenerational equality with respect to the environment (Gallon 1992:10). Additionally, anecdotal evidence suggests that ELCI provided, in part, the early intellectual foundation for sustainable development and presented these ideas to UNEP in a speech to the Governing Council in 1977 (Gallon 1992:27).

During the next year, UNEP, WWF, and IUCN jointly funded and worked on a document intended to provide an environmental policy blueprint for governments to adopt. The process proved to be controversial as IUCN focused on conservation while UNEP preferred a focus on environment and development.[30] The first draft of the World Conservation Strategy was published by IUCN in 1978. It makes no use of the term "sustainable development" although there is frequent reference to "sustainable use", particularly with respect to living resources such as mammals and fish. In fact, this report does not refer to development per se, although enacting conservation measures is frequently justified as serving the "wellbeing of humanity." The general disposition of this draft report toward the broader international political agenda is clearly articulated in the following:

> The conservation of wild living resources is only part of a wider field of endeavor, the conservation of all renewable natural resources, and, wider still, of general environmental policies aimed at the greatest lasting human well-being. This study is not specifically concerned with these wider problems (IUCN 1978).

This lack of interest in the broader problems of human wellbeing outside of conservation of living resources gives credence to ELCI's claim that sustainable development as a concept was formed during the 1976–1979 time frame and came into the final draft of the World Conservation Strategy through the direct intervention of UNEP in 1979 when representatives from UNEP (Pablo Bifani), IUCN (Robert Prescott-Allen), and ELCI (Gary Gallon) met at UNEP's offices in Nairobi (Gallon 1992:27). Gallon (1992:27) further claims that WCS adapted sustainable development directly from the 1977 Governing Council presentation. As a result of these discussions, sustainable development was incorporated into the World Conversation Strategy and released by the IUCN in 1980 (Omang 1980).

---

[29]The relationship between UNEP and NGOs was not one-sided as the ENGOs provided valuable scientific information and political activity in support of UNEP's agenda.

[30]In the 1980s, IUCN focused on preserving nature and natural resources for possible future use, while UNEP preferred to utilize natural resources efficiently in the present to improve the human environment.

Even after UNEP's intervention, the World Conservation Strategy did not truly embrace the integration of basic needs (Clansy 1992:27).

The report made a significant impact upon its release as it successfully repackaged conservation in a manner more consistent with the development priorities of the Third World. Additionally, IUCN over the course of the next several years began work on national conservation strategies that further strengthened the linkages between environment and development. This report was not the only UNEP funded activity dealing directly with environment and development. ELCI held a grant during the period 1978–1982 to produce resource kits for World Environment Day. The 1979 kit was oriented around the "International Year of the Child" and dealt specifically with environment and development (Gallon 1992:26).[31]

4.4 Sustainability Away from the UN

The previous sections implied that sustainable development came into existence during the late 1970s and early 1980s as a compromise between two different discourses, one promoting economic growth, termed here as the development discourse and the other insisting upon finite limits to growth, what Dryzek (2005: 16) called the limits discourse.

The development discourse is well known. Namely, states and markets should provide economic growth through wealth creation and employment. Environmental protection should not be allowed to halt economic growth. Consumption limits are acceptable as a personal choice, but not as a governmental policy goal.

The limits discourse promoted just exactly that, a limit to economic growth based upon the earth's ability to regenerate raw materials necessary to fuel the economy. Supporters of the limits discourse believe the earth exceeded its environmental capacity to support current economic consumption.

Academics typically credit the photographs of the earth taken by the Apollo astronauts for awakening this widespread environmental conscious. The "Spaceship Earth" picture of the blue planet on a black background caused people to think about the finite amount of natural resources during a period of prolonged economic growth beginning in the 1950s and ending with the oil shocks of the 1970s.

During the early 1970s the nascent environmental movement developed ideas about the survival of mankind in a world with limited natural resources. The Club of Rome, a group of elitist businessmen, scientists, and philanthropists came together to consider the potential impacts of overpopulation and overconsumption in 1968. The organization significantly framed much of the debate around environment and development in the 1970s, but did not necessarily engage directly with the UN bureaucracy to do so. Instead, the Club relied upon the quality of their ideas

[31]This kit was part of a broader UN effort to promote solutions to problems affecting children including malnutrition and lack of education.

and upon direct personal contacts, including overlapping membership between the political decision makers at UNEP and Club of Rome members.

The Club secured funding to commission a report by Massachusetts Institute of Technology researchers to examine various scenarios predicting when the raw materials used to support mankind would run out. The MIT team, led by Jay Forrester, applied state of the art computer models to examine the relationship between world population, industrialization, pollution, food production, and resource depletion. The report, Limits to Growth (Meadows et al. 1972), marked the first time a group of world public opinion leaders gave serious consideration to allowing natural resource conservation to stifle economic growth and development. Limits to Growth viewed all of humanity as a population with few capabilities towards self-direction and a world where elites with special expertise are needed to monitor, guide and direct the masses so as to avoid overconsumption of finite natural resources. Unsurprisingly, the policy conclusion stated that draconian measures by a supranational authoritarian government were necessary to halt economic growth and resource consumption in order to avoid economic collapse within the next 100 years, in this case, 2072.

However, a more vocal and widespread opposition encompassing prominent economists, research scientists, and political figures quickly organized a response and painted the book's authors as misguided doomsday prophets (Simon 1981; Simon and Kahn 1984). The arguments against Limits to Growth revolved around a myriad of technical and philosophical objections running the gamut from the inputs to the computer simulation to the future growth rates of pollution control technology. Opponents to the Limits to Growth philosophy argued that the computer inputs to systems modeling failed to consider mankind's capability to innovate both socially and technically to avoid a collapse of natural resources. Additionally, the Limits to Growth model treated all of humanity as a single system and neglected regional differences in the management of natural resources (Neurath 1984:68).

Equally important, powerful political groups devoted resources to mitigating the limits discourse. A broad coalition came together to voice opposition to specific tenets of the limits discourse. For example, the Roman Catholic Church along with Islamic fundamentalists traditionally oppose any measure dealing with population control. Communist states and other Marxist supporters vehemently derided the entire movement due to its elitist origins and capitalist backing. Indeed, the majority of the Third World countries opposed the limits discourse due to a deep seated fear that this would exacerbate the current geo-political inequalities in wealth distribution and permanently end their aspirations for economic advancement. At the end of the UNCHE in June 1972, the no growth rhetoric implied in the Limits to Growth was formally denounced in the Report of the UNCHE (UN 1972c).

Regardless of the arguments about the accuracy of the predictions, Limits to Growth permanently tied together environment and development at the international level. Limits to Growth appeared a scant 4 months before the first international environmental conference and directly impacted states' negotiating positions. Limits to Growth added a moral component to the global economic equality debate. Halting economic growth in the time frames specified in the Limits to Growth model had the net impact of permanently locking the underdeveloped countries into poverty. In doing so, Limits to Growth assisted in polarizing the North-South gap within global environmental affairs.

As a response to these criticisms, the Club of Rome accepted a second report on systems modeling entitled Mankind at the Turning Point by Mesarovic and Pestel (1974). This second report avoided the criticisms of Limits to Growth by halving the time frame of the study. This, in turn, allowed the authors and the Club of Rome more generally to avoid the heated debate about the assumptions surrounding natural resource depletion. The second report also increased the number of modeling inputs that in turn lessoned criticisms of the relationships between variables. Most importantly, Mankind at the Turning Point used regional representations for population, including political and social dimensions for resource management.

These changes in inputs allowed for a distinctive difference in policy conclusions and led directly to the "organic growth" discourse of the mid-1970s. Organic growth allowed for the increase in Third World development by reducing consumption in Western Europe and other countries. In other words, it distinguished between "good" and "bad" growth (Schmandt 2010: 43). Consequently, Mankind at the Turning Point found significant political support from the developing countries, especially as the report coincided with the UN negotiations on a NIEO.

The NIEO rhetoric dominated all facets of UN diplomacy from the 1974–1976 and continues to influence the politics of sustainable development. NIEO arose after the oil shocks of 1973 and caused the Third World to lose ground in closing the economic growth gap between the Western Europe and other countries and the Group of 77. Eastern Europe generally sided with the Third World by pointing out that the inequalities in economic distribution were a product of the capitalist West and did not occur in the Communist East. The NIEO rhetoric cast a gloomy pallor over international affairs and called on Western countries to self-sacrifice their economic growth in order to benefit their poorer neighbors to the South.

After Mankind at the Turning Point, the Club of Rome did not accept another environmental systems modeling report. That is not to say, however, that members of the club dropped environment and development from their agenda, but rather to assert that individual members of the Club of Rome continued working on the twin issues of environment and development outside of the Club of Rome. George Mitchell, owner of Mitchell Energy and a member of the Club of Rome, continued to foster an active debate about alternative limits to growth by establishing a series of conferences in The Woodlands, Texas. The conferences Mitchell sponsored brought together leading academic proponents of alternative economic models and allowed for the transformation of the limits discourse into the sustainability discourse.[32] This transformation did not occur smoothly, but rather meandered through several iterations of ideas, each idea moving successively closer to the definitions of

---

[32]Schmandt (2010) provides the most detailed account of the Woodlands Conferences in his biography of George Mitchell.

sustainability embedded within the formal UN report in 1987. The 1975 Woodlands Conference released the working papers under the title Alternatives to Growth (1977).

One of the seminal works within Alternatives to Growth was penned by Herman Daly, then an economist at Louisiana State University. He suggested that a transition from an economic growth model to a steady state economic model with both a constant population and a constant material base. (Schmandt 2010). Proponents of sustainable development today hold the steady state economic model as the ideal economy for the developed world. When combined with the portion of the limits rhetoric, or more correctly the mankind rhetoric, first world growth should occur in a steady state with a constant population and a constant material base while the third world should be allowed to grow in both areas until the development status of both groups of countries equalize. At that point, all countries should focus their efforts on maintaining a steady state population and raw material basis.

In addition to Daly's contributions of a steady state economic model, Donella Meadows advanced the discourse by coining the phrase "sustainable state." Mitchell, as reported by Schmandt (2010) claimed that he had the realization by the 1977 Woodlands Conference, that the discussion should not be centered around a sustainable state, but rather a sustainable society. This subtle shift in discourse allowed for the non-state actor and essentially shifted the boundaries of the subject area. Transnational corporations and individuals could potentially have an impact on what constitutes sustainability. Furthermore, transnational corporations were included in the ongoing debate about NIEO, although in a mostly unfavorable light. For the most part, however, transnational corporations did not play much of a role in orchestrating the sustainability discourse.

At least one link in the sustainability puzzle remains unresolved – namely how did the group meeting in the Woodlands transmit their ideas into the political process at the UN? It is not at all clear that this in fact happened, at least not deliberately after the beginnings of the Woodlands conference series. As stated previously, only one of the Woodlands conference proceedings was published, and that would not have incorporated work on the sustainable society. On the other hand, the alumni of the Woodlands conferences were all well-known and respected in their field, and it is likely that they also participated in a political process at both the national and international forums.

The second missing link is more problematic. The limits discourse reflects a zero growth philosophy that is distinctly different from sustainable development with its emphasis on equality of socio-economic justice. That is, there is a distinct difference between the meaning of sustainability in the Club of Rome context and sustainable development as used today and this difference in meaning suggests that separate groups worked simultaneously on defining the sustainable development discourse.

The events presented here correspond to and interact with the remainder of the case study in this chapter. Accordingly, the two parallel courses should be seen as complementary to each other; the UN does not operate apart from contemporary world events, but rather responds and interacts with the most promising political philosophies of the day.

4.5 Nairobi

At the request of the General Assembly, UNEP celebrated the tenth anniversary with a Session of Special Character of the Governing Council in Nairobi on May 3–12, 1982. The primary goals of Nairobi, or Stockholm + 10, were to review the Stockholm Action Plan and to give general guidance to UNEP for the next 10 years. Overall, the atmosphere stood in stark contrast to the first conference. Where Stockholm produced street festivals and a positive atmosphere for the future, Nairobi was negative and retrospective (Scharlin 1982; Selin and Linnér 2003). The downcast mood stemmed in part from the mandate for the event – to review progress of the Stockholm Action Plan though it was widely recognized that little concrete progress had been made during the 10 years since Stockholm.

The developing countries insistence on inclusion of human interactions with the environment was too much for other states. This cleavage resulted in both the fragmentation of the environmental agenda and the restriction of UNEP to providing the most basic of environmental activities, namely data collection and information sharing. The third factor in Nairobi's retrospective focus was the need to prove failure before expanding the role of environmental affairs. After all, one only fixes things that are broken.

Interestingly, *The World Environment 1972–1982*, one of the reports produced in UNEP Executive Director Mostafa Tolba's "year of the audit", documents the gradual shift in meaning within environmental affairs.[33] Given the negative atmosphere surrounding Nairobi, Tolba needed to walk a fine line. On the one hand, he had to show that governments' attitudes toward the environment had grown more positive, yet detail that states had never fully implemented the Stockholm Action Plan. On the other hand, Tolba had to acknowledge that the condition of the physical environment, the predominant concern in Stockholm, had given way to the interactions between humans and the environment, particularly the ways in which human activities adversely affect the environment.

The Nairobi conference serves as an oft-forgotten footnote in the history of sustainable development and it has become commonplace to forget this meeting in the annals of history (Goodland 1995; Lafferty and Meadowcroft 2000; Lipschutz 2004; Middleton and O'Keefe 2003; Redclift 1987; Rao 2000; Shabecoff 1996). However, this event provided important support for the WCED as the Governing Council recommended to the General Assembly "that it [establish] a voluntarily financed special commission to propose environmental strategies to the year 2000 and beyond" (UN 1982:1000). The motion moved through the UN system from UNEP to ECOSOC on July 29, 1982 (ESC 1982/56, para 7) and finally to the General Assembly on December 20 (GA A/Res/37/219). The General Assembly

[33] The other report of special interest is *From Stockholm to Nairobi*, a compilation of official statements by Heads of Delegations at UNCHE and the Special Session of the Governing Council of UNEP.

requested the Executive Director of UNEP to issue a report of the agenda and organizational structure for a commission after consulting with governments.

The Executive Director released the report in February 1983. It recommended that a committee of eminent persons prepare a document on environmental perspectives that he could use as a basis for future negotiations. The General Assembly approved the creation of the committee under the title Environmental Perspective to the Year 2000 and Beyond (A/RES 38/161). The resolution charged the Special Commission with articulating a long-term vision of environmental protection that would lead to greater cooperation among nations while at the same time taking into account the "different stages of economic and social development" of the various member-states. Tolba noted at the 8th UNEP Governing Council session that "the short- and medium-term preoccupations of the two groups [developed and developing countries] and their perception of immediate needs and constraints did not often coincide" (UNEP 2006). The General Assembly also requested that the Secretary-General of the UN, Pérez de Cuéllar, select a Chair and a Vice-Chair. After due consideration, de Cuéllar selected a Chair, Gro Harlem Bruntland (Norway) and a Vice-Chair, Mansour Khalid (Sudan), for this Special Commission. The General Assembly also instructed the Commission to "interact" with governments, scientists, other intergovernmental parties and the public. The General Assembly also noted that the Commission report would be a nonbinding report; that is, it would not have the status of mandatory international law.

Brundtland recruited a number of well-known government officials in environment and development including Maurice Strong (Canada), Jim MacNeill (Canada), Warren "Chip" Lindner (United States of America), and Sonny Ramphal (Guyana).[34] Interestingly, MacNeill, Ramphal, and Strong were, at the time, board members of the IIED, once headed by Barbara Ward. The Commission actively encouraged written reports from all countries. They also traveled to multiple locations to hear oral statements from all sectors of societies. At times, these requests for information conflicted with the policies of the states hosting public meetings. The Commission consolidated the information received and formulated a report based upon politically acceptable agreements between the individual members of the committee. IIED's Lloyd Timberlake and Richard Sandbrook served as prominent advisors to Brundtland, and Sandbrook (2003) later wrote that IIED provided initial drafts of three of the chapters in *Our Common Future*, and commented on the others.[35] As such, the report technically represents the viewpoints of its members acting in their individual capacity, but nevertheless is reflective of the current thinking of national governments. Timberlake writes (1989:122) that

[34]Jim MacNeill headed the environmental department of the OECD, Lindner was an executive director of the World Wildlife Federation, and Ramphal was the Secretary-General of the Commonwealth of Nations.

[35]Sandbrook does not specify which three chapters IIED initially penned. In a separate chapter, Satterthwaite (2003:125) identifies one of the three drafts as urban issues.

> Science was not the driving force in the making of the Brundtland Report. Although some premises were delivered by scientists, generally they were not attributed much weight. Nor were the main conclusions scientifically founded. ... The report is a *political* document, not a *scientific* one. In the hearings that were carried out in connection with the making of the report, testimonials made by 'ordinary' people were given much weight by the Commission. Generally, it appears that public opinion and fears are stronger driving forces for politicians than scientific reports.

Timberlake concludes that policy researchers, i.e. scientists who could communicate clearly with politicians had an important role to play in the implementation of the committee's report.

The committee, which had become known as the Brundtland Commission, after its chairperson, released its report *Our Common Future*, in 1987. It focused on the elaboration of sustainable development, a term coined in IUCN's 1980 report *World Conservation Strategy*. Sustainable development, as defined by the Brundtland report, is "development that meets the needs of the present without compromising the ability of future generations to meet their own needs" (WCED 1987). As such, sustainable development intertwines economic growth, social justice, and environmental protection. It seeks to end poverty, yet simultaneously to conserve the natural resources that undoubtedly will be consumed in the process. Further, sustainable development seeks to do so in a process that celebrates diversity and expands public participation (Cordonier Segger and Khalfan 2004:3).

More specifically, *Our Common Future* represented the Commission's work on three key objectives – (1) to examine the linkages between environment and development, (2) to outline issues where international cooperation might solve the twin problems of environmental degradation and underdevelopment, and (3) to raise awareness and commitment to sustainable development as part of the international political agenda. Yet the meaning envisioned by *Our Common Future* goes further than referring to an environmentally-friendly version of Western industrialization, or what Strong termed ecodevelopment.

Sustainable development, as defined by WCED, is subtly different from the Founex thesis. First, Founex did not refer to future needs. Second, Founex (and Stockholm) stated that developing countries might, or could, experience the environmental degradation of the more industrialized North while *Our Common Future* acknowledges that the untoward impacts of development have already become manifest. Third, many developing countries considered the more stringent environmental requirements of the North to be an "opportunity" to attract industry to the South, where the environmental damage would not be as severe. Several experts thought that the carrying capacity of the underdeveloped South was higher than the more industrialized North. That is, the South could absorb more easily than the North the incremental environmental damage from a new facility. Finally, the demands of the underdeveloped countries have changed over time as they have incorporated social concerns into pleas for financial assistance.

Political support for sustainable development, including the Brundtland Report, was not limited to governments and academics. Warren Lindner, the Secretary of the Brundtland Commission, established a non-profit organization, the Center for Our

Common Future.[36] The Center's original purpose was to promote the continuation of the debate and dialog prompted by the original report. In other words, the Center acted as a public relations agency whose job description was to create a "Brundtland constituency." The Center was at least moderately successful in identifying public support, particularly among NGOs including business associations, development organizations, and environmental groups. Particularly noteworthy in this effort was the distribution of a monthly newsletter, *Network '92,* whose worldwide circulation exceeded 100,000 in the lead-up to the Earth Summit in 1992.[37] As the process for Rio began to take shape, the Center "volunteered" to serve as the contact point for NGOs.

4.6 Conclusions

Ever since Sweden introduced a resolution calling for the Convention on Human Environment in 1968, the North and South have disagreed on the saliency and importance of this issue area. Ecodevelopment, which began as a compromise between the North's technically oriented environmental protectionism and the South's concern with industrialization and need to avoid a new neocolonialism, allowed the Stockholm Preparatory Committee to move out of the deadlock that negotiators initially encountered.

At the heart of the difference in negotiating agendas lie different meanings of "environment." For the North, environment represents the physical characteristics of the earth. The South takes a broader definition of the environment – the relationship between human beings and the physical environment. This broader Southern conceptualization more readily facilitates the linkage between environment and development and "the pollution of poverty."

The framing of environmental concerns as interlinked with development issues would eventually change the complexion of international environmental negotiations. It solved the problem of lack of interest and participation of the developing countries, but at the expense of greatly complicating the negotiating agenda. "Environment" is a term with no clear delimiters, allowing for this expansion into "the allocation of values within and between polities" (Kay and Skolnikoff 1972:472). With the rise of environmental protection, the South was able to expand a technical subject into a discussion of values that would ultimately lead to additional aid for development purposes. The WCED attempted to reconcile these different meanings under the auspices of sustainable development. This reconciliation was

[36] Lindner worked previously as the Deputy Director General of the World Wide Fund for Nature and at SOGENER, an energy investment company in Switzerland.

[37] Network'92 is archived at The Earth Summit NGOs CD-ROM, available at the Government Documents Department, Lamont Library, Harvard University.

superficially successful at the Earth Summit in 1992. However, the two tracks were in reality separated as the physical environment acquired a separate identity within the international regulatory system.

As such, proponents of sustainable development look outside of the traditional state-centric regime for norm acceptance. Thus, the role of NGOs has evolved from the "apolitical" status of those organizations during the pre-Stockholm era to a period in which the NGOs became embedded within the very structure of the state system at the center of global environmental politics. This move strengthened NGOs' capability to inject ideas into political debate and ultimately to shape the contours of global environmental governance.

References

Adams, W. M. (1990). *Green development environment and sustainability in the Third World.* London: Routledge.

Algeria. (1972). Statements by head of delegations – Algeria. In M. Tolba (Ed.), *Evolving environmental perceptions from Stockholm to Nairobi* (p. 114). London: Butterworths.

Artin, T. (1973). *Earth talk independent voices on the environment.* New York: Grossman Publishers.

Bartelmus, P. (2003). Dematerialization and capital maintenance: Two sides of the sustainability coin. *Ecological Economics, 46*(1), 61–81.

Bazell, R. J. (1972). Human environment conference: The rush for influence. *Science, 174*(4007), 390–391.

Berg, G. G. (1972). Stockholm conference. *Science, 177*(4051), 746.

Birnie, P. (1993). The UN and the environment. In A. Roberts & B. Kingsbury (Eds.), *United Nations, divided world the UN's role in international relations* (2nd ed., pp. 327–384). Oxford: Clarendon.

Boardman, R. (1981). *International organization and the conservation of nature.* Bloomington: Indiana University Press.

Caldwell, L. K. (1984). *International environmental policy. Emergence and dimensions.* Durham: Duke University Press.

Carson, R. (1962). *Silent spring.* New York: Houghton Mifflin.

Castro, J. A. (1972). Environment and development: The case of the developing countries. *International Organization, 26*(2), 401–416.

Clansy, D. (1992). *Sustainable development: The shift to a new paradigm.* Dissertation. University of Colorado.

Commoner, B. (1971). *The closing circle: Nature, man, technology.* New York: Knopf.

Cordonier Segger, M. C., & Khalfan, A. (2004). *Sustainable development law principles, practices and prospects.* Oxford: Oxford University Press.

Cross, N. (Ed.). (2003). *Evidence for hope: The search for sustainable development.* London: Earthscan.

Dryzek, J. S. (2005). *The politics of the earth: Environmental discourses* (2nd ed.). Oxford: Oxford University Press.

Eco. (1972a). *Dai Dong's declaration.* Bound as Stockholm conference Eco (Vol. 1). Stockholm, June 7, 1992.

Eco. (1972b). *A funny thing happened to the environment on its way to the forum.* Bound as Stockholm conference Eco (Vol. 1). Stockholm, June 14, 1972.

Eco. (1972c). *NGO come back, then to home.* Bound as Stockholm conference Eco (Vol. 1). Stockholm, June 14, 1972.

Eco. (1972d). *NGO go home*. Bound as Stockholm conference Eco (Vol. 1). Stockholm, June 13, 1972.
Engfeldt, L. (1973). The United Nations conference on the human environment – Some experiences. *International Organization, 27*(3), 393–412.
Falk, R. A. (1971). *This endangered planet: Prospects and proposals for human survival*. New York: Random House.
Feraru, A. T. (1974). Transnational political interests and the global environment. *International Organization, 28*(1), 31–60.
Feraru, A. T. (1985). Environmental actors. In K. A. Dahlberg, M. S. Soros, A. T. Feraru, J. E. Harf, & B. T. Trout (Eds.), *Environment and the global arena actors, values, policies and futures* (pp. 43–67). Durham: Duke University Press.
French, H. (1995). *Partnership for the planet: An environmental agenda for the United Nations*. Washington, DC: Worldwatch Institute.
Gallon, G. (1992). *The Environment Liaison Centre International: A short history*. Nairobi: Environment Liaison Centre International.
Gardner, R. N. (1972). The role of the UN in environmental problems. *International Organization, 26*(2), 237–254.
Goodland, R. (1995). The concept of environmental sustainability. *Annual Review of Ecology and Systematics, 26*, 1–24.
Gosovic, B. (1992). *The quest for world environmental cooperation: The case of the UN global environment monitoring system*. New York: Routledge.
Hamer, M. (2002). The filthy rich: How developed nations plotted to undermine global pollution controls. *New Scientist, 173*(2324), 7.
Herter, C. A., & Binder, J. E. (1993). *Role of the secretariat in multilateral negotiation the case of Maurice Strong and the 1972 UN Conference on the Human Environment*. Washington, DC: Foreign Policy Institute, The Paul H. Nitze School of Advanced International Studies, The John Hopkins University.
Hoggart, R. (1996). UNESCO and NGOs: A memoir. In P. Willetts (Ed.), *The conscience of the world* (pp. 98–115). Washington, DC: The Brookings Institution.
Holdgate, M. W., Kassas, M., & White, G. F. (1982). *The world environment 1972–1982: A report by the United Nations environment program*. Dublin: Tycooly International Publishing Limited.
Howarth, R. B. (1997). Defining sustainability: An overview. *Land Economics, 73*(4), 445–447.
Irwin, R. (2001). Posing global environmental problems from conservation to sustainable development. In D. Stevis & V. J. Assetto (Eds.), *The international political economy of the environment critical perspectives* (Vol. 12, pp. 15–38). London: Lynne Rienner.
IUCN. (1978). *First draft of a world conservation strategy*. Morges: IUCN.
Johnson, B. (1972). The United Nations' institutional response to Stockholm: A case study in the international politics of institutional change. *International Organization, 26*(2), 255–301.
Kay, D. A., & Skolnikoff, E. B. (1972). International institutions and the environmental crisis: A look ahead. *International Organization, 26*(2), 469–478.
Kennet, W. (1972). The Stockholm convention on the human environment. *International Affairs, 48*(1), 33–45.
Lafferty, W. M., & Meadowcroft, J. (2000). *Implementing sustainable development strategies and initiatives in high consumption societies*. Oxford: Oxford University Press.
Landsberg, H. H. (1972). Can Stockholm succeed? *Science, 176*(4036), 749.
Lipschutz, R. D. (2004). *Global environmental politics power, perspectives, and practice*. Washington, DC: Congressional Quarterly Press.
Luchins, D. (1977). *The United Nations conference on the human environment a case study of emerging political alignments, 1968–1972*. Dissertation. City University of New York.
Meadows, D. H., et al. (1972). *The limits to growth: A report for the club of Rome's project on the predicament of mankind*. New York: Universe Books.
Mesarovic, M., & Pestel, E. (1974). *Mankind at the turning point. The second report to the club of Rome*. New York: EP Dutton.

Middleton, N., & O'Keefe, P. (2003). *Rio plus ten politics, poverty and the environment*. London: Pluto Press.
Neurath, P. (1984). *From Malthus to the club of Rome and back: Problems of limits to growth, population control, and migrations*. New York: M. E. Sharpe.
Omang, J. (1980, March 6). Conservation strategy mapped by 30 nations. *Washington Post*, p. A13.
Palme, O. (1972). Speech of welcome by the Prime Minister of Sweden. In M. Tolba (Ed.), *Evolving environmental perceptions from Stockholm to Nairobi* (pp. 51–52). London: Butterworths.
Rao, P. K. (2000). *Sustainable development economics and policy*. Oxford: Blackwell Publishers.
Redclift, M. (1987). *Sustainable development exploring the contradictions*. New York: Methuen.
Rowland, W. (1973). *The plot to save the world*. Toronto: Clarke, Irwin & Company Limited.
Sandbrook, R. (2003). Change that works, sometimes. In N. Cross (Ed.), *Evidence for hope the search for sustainable development* (pp. 60–74). London: Earthscan.
Satterthwaite, D. (2003). Setting an urban agenda: Human settlements and IIED-América Latina. In N. Cross (Ed.), *Evidence for hope the search for sustainable development* (pp. 122–146). London: Earthscan.
Scharlin, P. (1982). The United Nations and the environment: After three decades of concern, progress is still slow. *Ambio, 11*(1), 26–29.
Schmandt, J. (2010). *George P. Mitchell and the idea of sustainability*. College Station: Texas A&M University Press.
Selin, H., & Linnér, B. -O. (2003). *The quest for global sustainability: International efforts on linking environment and development* (Working Paper No. 5). Cambridge: Science, Environment and Development Group, Center for International Development, Harvard University.
Shabecoff, P. (1996). *A new name for peace international environmentalism, sustainable development, and democracy*. Hanover: University Press of New England.
Simon, J. (1981). *The ultimate resource*. Princeton: Princeton University Press.
Simon, J., & Kahn, H. (1984). *The resourceful earth: A response to the global 2000 report*. New York: Basil Blackwell.
Spiro, P. (1994). New global communities: Nongovernmental organizations in international decision-making institutions. *The Washington Quarterly, 18*(1), 45–56.
Stone, P. B. (1973). *Did we save the earth at Stockholm?* London: Earth Island Limited.
Strong, M. F. (1972). The United Nations and the environment. *International Organization, 26*(2), 169–172.
Strong, M. F. (2000). *Where on earth are we going?* Toronto: Alfred A. Knopf Canada.
Timberlake, L. (1989). The role of scientific knowledge in drawing up the Brundtland report. In S. Andersen & W. Ostreng (Eds.), *International resource management: The role of science and politics* (pp. 117–123). New York: Belhaven Press.
Tinker, J. (1975). Cocoyoc revisited. *New Scientist, 67*(964), 480–483.
Tinker, J. (2003). Earthscan – Panos: Favourite Son, Via Cuckoo-in-the-Nest, to Friendly Rival. In N. Cross (Ed.), *Evidence for hope the search for sustainable development* (pp. 40–59). London: Earthscan.
UN. (1968). *ECOSOC resolution 1346 (XLV)*. July 30, 1968.
UN. (1969). *General assembly resolution 2581*. Retrieved from http://daccess-dds-ny.un.org/doc/RESOLUTION/GEN/NR0/257/15/IMG/NR025715.pdf?OpenElement
UN. (1972a). *Development and environment. Report and working papers of a panel of experts convened by the secretary-general of the United Nations conference on the human environment (Founex, Switzerland, June 4–12, 1971)*. Paris: United Nations.
UN. (1972b). *General assembly resolution 2997 (XXVII)*. Retrieved from http://daccess-dds-ny.un.org/doc/RESOLUTION/GEN/NR0/270/27/IMG/NR027027.pdf?OpenElement
UN. (1972c). *Report of the United Nations conference on the human environment* (A/CONF.48/14/Rev.1). Retrieved from http://www.un-documents.net/unche.htm
UN. (1982). *Yearbook of the United Nations*. New York: United Nations.
UNEP. (2006). *8th Governing council session*. Section II. General update. Retrieved from http://www.unep.org/Documents.Multilingual/Default.asp?DocumentID=62&ArticleID=472&l=en

UNESCO. (2005). *About UNESCO*. Retrieved from http://portal.unesco.org/en/ev.php-URL_ID=3328&URL_DO=DO_TOPIC&URL_SECTION=201.html
Ward, B., & Dubos, R. (1972). *Only one earth: The care and maintenance of a small planet*. New York: Norton.
WCED (World Commission on Environment and Development). (1987). *Our common future*. New York: Oxford University Press.
Willetts, P. (1989). The pattern of conferences. In P. Taylor & A. J. R. Groom (Eds.), *Global issues in the United Nations' framework* (pp. 35–63). New York: St. Martin's Press.

Chapter 5
Reflections

Abstract This chapter pulls together the key elements in the book to present an analysis of nongovernmental organizations' influence during the creation of the sustainable development discourse. The chapter forms three key propositions based upon the case study. The key propositions give an overview of the theoretical importance of this research, namely, that NGOs successfully bargained their special expertise into a seemingly permanent symbiotic relationship with UNEP. The next section assesses the explanatory power of current global environmental governance theory and concludes that existing approaches explain some facets of this case study. However, current theories generally fail to consider NGOs role as a policy implementer and also fail to assess NGOs influence while engaging in this activity. The fourth section fills this gap in scholarly theories. Similarly, the fifth section reviews conditions impacting NGO influence. This chapter also concludes that NGOs influence did not begin at the end of the Cold War.

Keywords Goal attainment • Nongovernmental Organizations (NGOs) • Influence • Conditioning factors • Rational moral authority • Cold War

5.1 Introduction

Chapter 4 presented a historic narrative detailing the rise of sustainable development. The insights this case study offers has merit in the present because the international system that exists today is a function of the decisions and non-decisions of yesterday. In certain respects, the basic structure of international environmental governance established at Stockholm in 1972 has not been significantly altered. This system has expanded, particularly after the Rio de Janeiro Earth Summit in 1992. Additionally, the patterns of engagement with respect to the development of new discourses have remained relatively constant in that change has predominantly occurred through small incremental steps rather than through large strides, although dramatic breakthroughs have occurred.

A.E. Egelston, *Sustainable Development*, DOI 10.1007/978-94-007-4878-1_5,

This chapter analyzes the case study with the theories presented in Chap. 4 in order to answer the research question. As a result of this research, an additional theoretical quandary emerged – is the end of the Cold War an explanatory factor for the increase in political activity and influence of NGOs.

5.2 Three Key Propositions

The short synopsis presented above leads to three generalized principles about NGO influence. As stated previously, these are not the only identifiable propositions from this case study, but instead are those which lead directly to an examination of the role, function, and political influence of NGOs.

Proposition 1: *Although not subject to scholarly research at the time, NGOs made significant contributions to the UN beginning with UNCHE in the 1970s, including the deliberate institutionalization of NGOs' activities and special expertise.* The sheer numbers of NGOs attending Stockholm, while small compared to the later meetings, was unprecedented and extraordinary for the 1970s. NGOs had the right to observe the UN system since its beginnings in 1945 based upon their active participation in the failed League of Nations. However, the level of political activity of NGOs in the UN system was greatly reduced from the League of Nations. NGOs had lost status and had become apolitical observers, rather than political actors with agency.

The decisions made by Strong during the Stockholm process – most notably the expansion of the "right" of observation to any NGO with a special interest in the environment – significantly broadened the number and kind of NGOs participating at a UN conference. At a time when a mere 519 NGOs were registered with ECOSOC, roughly 300 formally attended the UN meeting, along with a second set of non-registered observers at the Hog Farm (Willetts 1996:38). The expansion of participation of NGOs to those groups not familiar with the UN (and member states) expectations undoubtedly led to the expansion of the political activities of NGOs. While "new" NGOs were not the only groups that engaged in lobbying at Stockholm, they were certainly more likely than their older predecessors to do so.

NGOs activities at Stockholm, and for all conferences afterwards, essentially fall into three categories: observing, lobbying, and project implementation. Within these three categories, past precedence indicated that NGOs would primarily observe the conference. Although many organizations undoubtedly went to Stockholm to get information about the ongoing negotiations, NGOs, once at the conference, went considerably further than the narrow, nonpolitical role envisioned by the UN charter. The lack of mention of NGOs' political activity within the secondary literature of the time indicates that scholars did not expect NGOs to take an active role in lobbying and project implementation. After careful examination of the existing conference records and contemporary writings, this lack of expectation does not match the evidence from the conference participants, including NGOs themselves, who were concerned not only with the formal rights of observation, but also with both lobbying and project implementation.

NGOs' lobbying began during the Stockholm conference preparatory meetings, more specifically after Strong became the second Conference Secretary-General in 1970. The methods used by NGOs included submitting reports to the conference secretariat, drafting initial negotiating documents, publishing a conference newsletter, issuing joint statements, hosting side events, and publishing alternative treaties. In all likelihood the active contributions of NGOs, particularly ENGOs, to the Stockholm Conference raised world opinion in favor of government's pro-active engagement on environmental issues, including the issuance of statements of principles, as well as concrete action plans. This role of pressuring governments into action continues today, although with varying degrees of success.

NGOs also were active in implementation of the conference recommendations, as they designed an NGO infrastructure for the future that included networking with other NGOs, as well as interfacing with the newly created UNEP system. NGOs published a directory of NGOs at the Stockholm Conference, created a new UNEP watchdog INASEN, and held follow-up meetings to the Stockholm conference to continue discussions about environmental issues as well as to coordinate activities with the newly formed UNEP. Additionally, NGOs transformed their identities in the aftermath of the Stockholm conference through the incorporation of developmental considerations, as exemplified by the alteration of the IIEA into the IIED. In short, just as states made agreements to create the international environmental governance infrastructure, likewise NGOs also made plans to monitor, lobby and interface with this new infrastructure. Further, there was no lag time between the creation of UNEP and the beginning of NGOs' political engagement, including the establishment of the UNEP–NGO consultancy arrangements.

Despite these activities, government officials and members of the UN bureaucracy clung to the idea that NGOs served as observers, sources of information and mobilizers of public opinion. By this time, however, NGOs saw themselves differently and came prepared with substantial lobbying plans and agendas, not only for government and UN officials, but also for interacting with other non-state actors including BINGOs and transnational corporations (TNCs).

NGOs enjoyed a symbiotic relationship with UNEP, providing political opportunities and much needed financial resources. As the UN found its funding to be in continual jeopardy (along with the severe resource constraints of the majority of its members), NGOs increasingly served as an alternative resource provider rather than the "rich" UN member states. NGOs provided such necessities as access to specialized knowledge and free labor while working to establish public opinion in favor of progressive outcomes that might not have been readily forthcoming. In exchange for their services, the UN provided to NGOs not only a reason for their existence, but also limited financial support for programs. Consequently, the political bargain struck in the 1970s linked the UN to a small, specific number of locales.

Proposition 2: *The structure of international governance was designed deliberately to take into account NGOs activities and special expertise.* It was not a system of piecemeal activities and opportunistic decisions and events, although these things undoubtedly helped NGOs find a niche with both UN officials and national governments. UNEP and NGOs enjoyed a symbiotic relationship with UNEP providing political opportunities and much needed financial resources. As the UN found its funding to be in continual jeopardy (along with the severe resource constraints of the majority of its members), NGOs increasingly serve as an alternative resource provider than the "rich" UN member states, particularly with respect to such necessities as access to specialized knowledge and free labor working to establish public opinion in favor of progressive outcomes that might not have been readily forthcoming. In exchange for their services, the UN provided to NGOs not only a reason for their existence, but also limited financial support for programs. Consequently, the political bargain struck in the 1970s linked the UN to a small, specific number of locales.

While the structure of international governance was one of the key issues at the Rio Earth Summit and was discussed in the run-up to the Johannesburg conference, the interface between the UN system and NGOs changed little in practice. The net effect of the addition of the CSD Roster was to codify and classify an easily identifiable, politically active subgroup of NGOs. The addition of the CSD to the UN system adopted the working practices of the UNEP General Council Sessions – that NGOs were invited to make presentations. The CSD did not establish new precedence in that NGOs were now allowed to make presentations to a standing body of the UN, that practice started with the UNEP General Council Sessions in the 1970s. The changes made at the Earth Summit merely expanded the working practices of UNEP into a new organization, one whose mission and mandate have not been clearly defined from UNEP in any respect. In retrospect, the Earth Summit made apparent the rapid expansion of the political bargain that NGOs and the UN had developed some two decades earlier.

Proposition 3: *Non-state actors that had expertise in development economics (and the development agenda, more generally) also influenced the outcome of sustainable development.* Environmental politics has never been solely about the physical characteristics of the environment, despite Northern insistence on this position, but also about quality of life and the creation of wealth, along with the maintenance of natural resources. While NGOs with scientific knowledge and expertise provided invaluable support to the UN secretariat, non-state actors with economic expertise moved sustainable development forward at critical points. The political blend of environment and development agendas did not occur within the nation-state system, but rather from a network of like-minded NGOs and individuals. This network slowly conceptualized sustainable development beginning in 1974 with the Cocoyoc Declaration and continued to build on this framework until its adoption in the Brundtland Commission report in 1987.

This pattern of mixing environmental characteristics and development economics alleviated political opposition to the addition of the environment on the international agenda in the 1970s most especially at Founex and Cocoyoc. It is not an overstatement to say that sustainable development owes more to the developmental experts working on this issue in the 1970s than to the environmental activists during the same time. When environmental activists pursued items such as the conservation of natural resources, developmental economists merged this item with their own agenda adding that natural resources should be well-used in order to promote both economic growth and social justice.

In fact, environmental groups resisted this merged agenda during the late 1970s. The first draft of the World Conservation Strategy in 1978 rejected the concepts of "basic needs". It was only after ELCI, a third world NGO, successfully involved UNEP (& presumably the withdrawal of UNEP funding) that sustainable development made its way into a major international policy document. Even so, IUCN unsurprising emphasized environmental protection over economic growth, a frame alignment that remained intact for the next decade, until other third world NGOs such as Third World Network successfully shifted the sustainable development discourse at the Earth Summit and other later events.

5.3 Global Governance Theory Revisited

Chapter 3 began by recognizing that NGOs are one important component of global civil society. They contribute to stability and order within international life, but definitional mismatches between global civil society and NGOs working inside the UN make these theories difficult to use in this analysis. Regime theory narrows the focus of the political arena to UN institutions, but these theories traditionally limit NGOs role to providing legitimacy to existing regimes. Global environmental governance seeks to meld these two approaches so that global civil society impacts not only regimes, but also the broader processes of environmental governance, including discourses. These theories may provide an avenue for assessing NGOs' influence upon sustainable development.

Chapter 3 states that NGOs form into multiple transnational advocacy networks around discourses representing subtle, but important differences in the hierarchy of norms. Differences in the norm varies by political group. Some groups support limits to economic growth, others argue for increased economic growth, while others emphasize resource conservation instead. This polarity gradually widened in the years after the Stockholm conference in 1972 as the values underpinning sustainable development diverged from traditional environmental conservation and stewardship.

Further this research theorized these groups create social change by linking the local with the global through the use of specialized knowledge about environmental damage combined with their moral rational authority to identify solutions to this problem. Princen and Finger (1994:38) theorized that NGOs in the international system operated by linking the local to the global while fulfilling the role of an agent promoting social learning.

> NGO bargaining with other international actors, then, is one dimension of the NGO niche in international environmental politics. A second dimension is spatial. That is, NGOs position themselves within both top-down and bottom-up approaches to international environmental policy-making by attempting to link local needs with the challenges of the global ecological crisis.

The specialized knowledge that the NGOs had acquired could be deemed symptomatic of a "bottom-up" process, that NGOs learned about their damaged environment locally, then forced their agenda up towards the international level, and there is an element of truth in this argument (Korten 1990). Yet one criteria of a "bottom-up" process – success as measured by improvements in the local environment – never enters into serious consideration at the international level. Indeed, NGOs' haste to build new networks and to interface directly with UNEP in and of itself provides evidence against the "bottom-up" process.

Alger (1999:395) stated "that it would be good to know when, and why, influence is moving from the bottom-up, and when, and why, from the top down." The theoretical propositions shed some insight into this question. Structural political influence moves from the top-down, while agency political influence moves from the bottom-up. The "why" answer stems from the very nature of NGOs themselves – their dual location in both structure and individual agent. In this case, then, forces are at work that are neither strictly speaking top-down nor bottom-up in theory, but rather some combination of the two approaches.

In providing the local-global link, NGOs would necessarily be involved in both top-down and bottom-up processes of international governance. This theory comes closer to explaining the influence of NGOs in sustainable development, in that it describes a political bargain between NGOs and other elements of the international system. The political bargain allows access to the political decision-making processes in exchange for use of NGOs specialized knowledge and public relations skills.

One problem with this theory is the implication that two sets of entirely independent networks or agents exist – one for the NGOs and another for the UN machinery and that these two agents engage in political bargaining. Because this case study deals with, in part, the creation of the global environmental governance system, it does not exactly match the preconditions specified by Princen and Finger. Two entirely independent networks of actors did not, strictly speaking, exist; that is, NGOs are not lumped into one network with all of the UN machinery in the other network. The UN machinery in this case also participated in all of these networks. Additionally, these networks share related, but not identical, goals. Furthermore, procedural rules of access impacted both groups equally. From the onset, the innermost circle of NGOs, those with behind the scenes access to the UN, had their origins in part in UN bodies such as UNESCO, or through close personal contact with high-ranking UN officials. In other words, there is a structural dependency factor, either from shared material resources or from overlapping personnel that cannot be captured by the local-global linkage theory.

Kütting and Lipschutz (2009) remind theorists to consider knowledge as a function of the structural location of the knowledge provider; that is, knowledge about environmental destruction and ideas to solve this problem may look different depending upon the NGOs orientation in place and space. Consequently, the North-South divide that characterized (and continues to color) UN politics also impacted the sustainable development discourse. NGOs working locally in the G-77 and in China tended to view environmental problems as part of a broader problematique that also included social justice and the need for continued economic development. In contrast, NGOs with roots in the more developed North tended to use a narrower definition of environmental degradation.

However, provision of specialized knowledge is not the only function that NGOs perform. ELCI's role in promoting sustainable development moves beyond the function of providing specialized knowledge into the realm of norm development. Keck and Sikkink (1998:35) point out that "Norms and practices are mutually constitutive – norms have power in, and because of, what people do." When ELCI made a presentation on sustainable development to UNEP at the 1977 Governing Council meeting, the content of that presentation took on a life of its own. UNEP rapidly adopted the ideas. Further, ELCI and UNEP transferred sustainable development to IUCN in 1979, when they sought to promote sustainable development as a key idea in the 1980 World Conservation Strategy. In doing so, they hoped to not only question current practice, but also to constrain the behaviors of others in the future by eliminating environmentally destructive actions.

While some NGOs clearly viewed themselves as part of a broader social movement such as a transnational advocacy network, NGOs' willingness to implement public policy decisions moves them beyond the realm of social movement theory. NGOs should be properly categorized as a unique group unto themselves –independent actors capable of producing new knowledge, stimulating social learning, and acting and reacting to the social, cultural and physical phenomenon that comprise global environmental governance.

As independent actors with freedom from the formalities of governments and the rule of law, NGOs have inherent within their internal nature the freedom and flexibility necessary to engage in public policy innovation and experimentation. NGOs recognized a need for innovation and multiplication of policy programs, including structural opportunities and brainstorming sessions, particularly where normative values enter into the political realm.

Consequently, existing theoretical approaches should uphold NGOs as a prominent actor within global environmental governance. NGOs, with their rational moral authority, are uniquely qualified to implement new policy programs based upon their combination of specialized knowledge and ability to generate public support for solutions to these problems. While many of these ideas about new norms or new policies were originally generated outside of the hallways of the UN, the consultancy relationship between UNEP and NGOs created an infrastructure that allows these ideas to move freely between these two actors. The absence of the role of an individual state is particularly noteworthy with respect to innovation and experimentation. While states clearly have a great deal of impact on regime formation through their dominating role at the UN, no individual state took up the challenge of positioning itself as a hegemonic force within the sustainable development discourse.

5.4 NGO Goals

If NGOs are an agent who matters, they should have, under certain circumstances, political influence stemming from their rational moral authority and that this authority can be converted into influence when NGOs meet their goals. Chapter 4 presented a case study highlighting major events leading up to the Brundtland Commission report – Our Common Future. The analysis begins with constructing a list of NGOs activities during this time and classified the activity by the goal it sought to accomplish. The next part of the examination of NGO influence evaluates whether or not NGOs achieved these goals.

5.4.1 Goal Achievement

NGOs undoubtedly achieved their first goal, gathering information about the current state of the environment. They also gathered information about states' proposals to halt or reverse environmental degradation. Both networks engaged in this activity. However, gathering information about environmental degradation and political solutions need not lead to NGOs influencing the sustainable development discourse. In other words, NGO influence does not automatically stem solely from observing international environmental negotiations and suggesting solutions. Feraru (1974) recognized that presence at the table does not equate to influence with political decision-makers when she stated:

> Nongovernmental associations with diffuse humanitarian objectives, no matter how worthy, are likely to be heard politely and then ignored – unless they can demonstrate ability to inform and mobilize visible support for UNEP. Contacts with the more radical eco-activists will probably be avoided by the secretariat if possible, and such groups' lack of familiarity with the protocols of conventional intergovernmental procedures tends to make them ineffective in reaching the secretariat and governments through the usual channels.

Conference secretariat, the UN bureaucracy, and government officials more generally, have the option of attending meetings, even actively engaging in discussion with NGOs, but more often than not, these officials ignore NGO advice in decision-making processes. Being at the table and engaging in dialogue, is not, in and of itself, political influence in the sense that the agent contributed successfully to treaty outcomes. That is not to say that gaining information about the conference or the underlying state of the environment is an unimportant goal. On the contrary, the international conference legitimized the individual NGO in the eyes of their members.

Receiving information from the international political arena constitutes a low level of influence in that the NGO must use their rational moral authority to determine what ought to happen as a result of this information. In essence, NGOs screen information through their moral filters prior to supporting policies, norms, or beliefs. NGOs that successfully mobilize visible support for either the political process or for the outcomes of the treaties are more likely to have influence in the future. Such is the case for the sustainable development discourse. NGOs successfully mobilized public opinion in support of the Stockholm conference by educating their members about the process and possible political outcomes. In doing so, the conference secretariat probably lowered NGO criticism of the conference proceedings and outcomes as NGOs would have more difficulty speaking out against documents and proceedings they assisted in creating.

5.4.2 *Providing Expert Assistance*

The second goal stated that NGOs have influence when they provide expert assistance to decision-makers and that they were highly influential when they achieved this goal across both networks. Princen and Finger (1994:11) describe NGOs in functionalist terms as actors who "can research, publicize, expose and monitor environmental trends with little fear of offending constituencies or losing customers." In doing so, they provide specialized knowledge to governments about the state of the environment as well as about activities that cause environmental degradation. IUCN's advice to the Stockholm conference secretariat serves as one example of how NGOs fulfilled this function. This organization, along with several others, used their specialized knowledge to issue reports on pollutant monitoring, international registry for chemical compounds, and implications of "man-made" ecosystems. Further, NGOs provided early drafts of reports and recommendations to the Stockholm conference secretariat; thus, allowing them to directly influence the formal outcomes of this meeting.

The UN secretariat, along with most member states, had a need for the specialized knowledge and political support of NGOs that could not be met easily through the expenditure of state (or UN) resources. Reimann (2006:46) was correct in her use of political opportunity structure, especially the two components of resource mobilization and political access as crucial variables.

> As international institutions and regimes have expanded to handle new global issues, they have increasingly promoted NGOs as their service providers and advocates, and in the past two decades an explosion of new international opportunities for funding and participation of NGOs has created a structural environment highly conducive to NGO growth.

In this sense, NGOs provision of "free" knowledge functioned as an alternative to state-sponsored resources, although there was never an attempt by NGOs to replace states, nor would the UN find it desirable to do so. UNEP's acceptance of this knowledge came with an implicit political bargain – NGOs gained high level access to decision-makers within UNEP. This relationship, or rather, the series of relationships, over time provided the neural pathways that directly shaped the sustainable development discourse.

Further, the expert assistance NGOs rendered was not limited to environmental knowledge or to the environmental conservation network. NGOs with expertise in development economics (and the development agenda, more generally) also influenced the normative values of sustainable development. Environmental politics has never been solely about the physical characteristics of the environment, despite Northern insistence on this position, but also about quality of life and the creation of wealth, along with the maintenance of natural resources. While NGOs with scientific knowledge and expertise provided invaluable support to the UN secretariat, NGOs with economic expertise moved sustainable development forward at critical points. The political blend of environment and development agendas did not occur within the nation-state system, but rather from a network of like-minded NGOs and individuals. This network slowly conceptualized sustainable development beginning in 1974 with the Cocoyoc Declaration and continued to build on this framework until its adoption in the Brundtland Commission report in 1987.

5.4.3 Lobbying Decision Makers

The third goal of NGOs was to lobby decision-makers as well as other participants. In this analysis, lobbying is narrowly defined as including only occasions when NGOs ask decision-makers to take a specific action on a specific item such as a treaty text or recommendations. For example, NGOs achieved one of their most notable lobbying successes when they convinced states to insert the moratorium on whaling into the Stockholm Action Plan. However, NGOs also had many public failures in attempting to broadly lobby policymakers. Requests for Stockholm to also address limits to growth, population control, and unequal distribution of resources failed. Furthermore, the Dai Dong conference failed to reach the same level of agreement as the official conference due to the wide variety of underlying beliefs and norms among conference participants.

NGOs seeking to influence national governments after Stockholm also achieved mixed results. NGOs lobbied national governments in support of UNEP and pressured states to continue to support the moratorium on whaling. States provided some funding for UNEP; however, the organization is routinely underfunded and there is no clear evidence that NGOs impacted the level of funding UNEP received from states. Likewise, the IWC eventually agreed to a moratorium on whaling, although this success was not realized until 1982. NGO efforts to ban commercial whaling were strengthened when like-minded states joined the IWC and the inclusion of these states undoubtedly assisted NGOs. Consequently, the assessment is that NGOs achieved a medium level of influence on these two issues in that NGOs achieved their goals with some observable change in norms and outcomes; however, it is not clear that the result directly depended on NGOs lobbying efforts.

There were also failures. NGOs' attempts to halt offshore drilling failed. Similarly, NGOs also failed to convince states to fully implement the Stockholm Action Plan. States never seriously considered banning offshore oil drilling and the practice continues worldwide today. Likewise, states ignored NGOs' pleas to fully implement the Stockholm Action Plan leading UNEP to conclude in 1982 that state's attitudes towards environmental affairs shifted in favor towards protection, but these shifts in attitudes did not lead to the fulfillment of the Stockholm Action Plan. NGOs did not have influence on these two issues. Thus, because of the disparity between successes and failures, it is difficult to determine how much influence NGOs exert while lobbying decision-makers. The causes of successes and failures appear to be unrelated to NGOs rational moral authority.

5.4.4 Representing the Organization

NGOs achieved their fourth goal – representing the organization's general views about values, norms and beliefs to other political players. The question, then, is whether or not any meaningful changes in the sustainable development discourse occurred due to these activities. During the Stockholm conference, NGOs were generally viewed as an interesting sideshow, with little to offer in terms of concrete policies and recommendations, with a few notable exceptions – Margaret Mead's speech to the plenary being one of those exceptions.

After the Stockholm conference, NGOs attempts at coalition building to interface with UNEP largely failed, although individual NGOs created for this purpose reconstituted their identity and corresponding mission statement. NGOs agreed to work together for the purposes of working with UNEP in the future and created a new NGO, INASEN for this purpose. INASEN, however, became embroiled in a legal controversy with UNEP's host country, Kenya. INASEN responded by transforming itself into the politically powerful ELCI. ELCI, in turn, generated significant efforts to creating the sustainable development discourse by articulating the principles. ELCI claims they introduced intergenerational equality into an ongoing debate at UNEP about irrational use of natural resources and coined the phrase sustainable development during the 1976–1979 time frame.

Consequently, NGOs exerted a high level of influence when achieving this goal. Despite Feraru's suggestion that this goal might not be important, NGO success led to a significant, albeit unintended consequence for the sustainable development discourse – forming ELCI. NGOs' successes in achieving this goal were obscured by two factors. First, NGOs rarely seek to highlight a general statement of norms without at least one specific policy outcome. This results in significant difficulty in distinguishing representing the organization from lobbying decision-makers. Second, Feraru conducted her assessment of NGOs' goals at Stockholm prior to INASEN's transformation into ELCI and it would not be possible to assess the success or failure of NGOs ability to represent the organization at that time.

5.4.5 Implementing Policy

The fifth NGO goal deals with policy implementation, including the adaptation of treaty recommendations and requirements. NGOs had great success and high influence in policy implementation during the time period covered by this case study. The Stockholm Action Plan (Recommendations 45, 95, and 98) generally requested that NGOs continue to provide technical and financial support to generate and disseminate information about the state of the environment and to mobilize public support for environmental protection.

NGOs' plans, however, went beyond what states were willing to cement in writing. Consequently, NGOs designed their own infrastructure that included networking with other NGOs, as well as interfacing with the newly created UNEP system. NGOs published a directory of NGOs at the Stockholm Conference and created a new UNEP watchdog. In short, just as states made agreements to create the international environmental governance infrastructure, NGOs also made plans to monitor, lobby and interface with this new infrastructure. Further, there was no lag time between the creation of UNEP and the beginning of NGOs' political engagement, including the establishment of the UNEP–NGO consultancy arrangements.

NGOs' influence formed concurrent with the establishment of the international environmental structure (i.e. UNEP), in a symbiotic relationship with that organization, and indeed, with the rest of the institutional structure for global environmental governance. NGOs continued to mobilize support for UNEP in the early 1970s that led to UNEP funding these campaigns. While the monetary value of these funds were small even by UNEP standards, they were nevertheless vital in shaping the NGO-UNEP relationship as well as embedding NGOs within the formal structures of global environmental politics. They are simultaneously an integral part of the international institutional structure as well as an individual actor. Giddens (1984:84) wrote:

> The positioning of actors in the regions of their daily time-space paths, of course, is their simultaneous positioning within the broader regionalization of societal totalities and within intersocietal systems whose broadcast span is convergent with the geopolitical distribution of social systems on a global scale.

NGOs are permanently locked into a second-class structural status as "non-state", yet retain the flexibility as agent to form into powerful coalitions, that have, at times, focused sufficient influence so as to create new agendas and new governing philosophies within the international system.

The provision of specialized knowledge and the creation of new norms are both indirect tools, in that they do not change the state of the environment. Rather, environmental improvement requires action by a third party. NGOs' use of project implementation as a political tool has increased the legitimacy of these actors, not only through their ability to deliver concrete results but also through the increase of specialized knowledge that only comes from hands-on experience.

NGOs' use of project implementation as a political tool has increased the probability that environmental protection will improve. NGOs move from indirect advocates to direct sponsors of programs, including pilot programs, increasing the likelihood that the projects will yield measurable benefits to the local environment because these groups are less likely to accept changes in the political system as a proxy for effective global environmental governance. Yet this move into project implementation is not new for NGOs. In the time frame of the Stockholm conference and the years following, NGOs engaged in the same project that states took up – of creating new relationships with the new international structures under construction. Further, NGOs had a need to build these relationships so that they were not clearly visible from the outside looking in, resulting in the quiet consultancy situation that continues to exist today. As sustainable development moved into different phases, NGOs' projects changed from items that dealt with defining principles and norms into concrete actions easily replicated by others.

Additionally, NGOs' move into project implementation solidified their identity as an independent actor separate from political lobbyist, or scientific advisor. NGOs are neither a mere lobbying force akin to a political interest group at the international level, nor are they an alternative to a state. NGOs within the UN system do more than seek to influence the decisions made by other actors. At the same time, they are also not merely an alternative knowledge provider to national governments and international bureaucrats. Further, NGOs are not a curious hybrid of these two functions that have merged into one organization under pressure to consolidate functions. It is perhaps not an oversimplification to think of NGOs as akin to resident aliens –independent actors living within a specific system, but yet without the full rights of citizens, despite a wholehearted attempt on their behalf to adhere to the requirements of international governance. There is no route by which NGOs may become full-fledged members of the UN system, nor is it necessary for NGOs to do so in order to continue to expand their political influence.

5.4.6 Keeping Issues Alive

The sixth accomplishment of NGOs deals with their ability to mobilize support in order to keep issues alive over long periods of time. In the case of sustainable development fully 20 years elapsed between the Stockholm Conference on the Human Environment and the Rio de Janeiro Conference on Environment and Development. In the intervening years, NGOs participated in numerous smaller conferences and symposiums in order to give voice to the norms and values of sustainable development. Keeping an item on the agenda for 20 years is no small feat for any political actor, whether it be a state, intergovernmental organization, or NGO.

Support for UNEP came from not only existing NGOs such as IUCN, but also from NGOs fashioned solely for the purposes of supporting UNEP policies. NGOs participating at Stockholm set in motion a plan to coordinate and interface with UNEP, leading to the establishment of ELCI. Further evidence of the close relationship between UNEP and NGOs includes Barbara Ward's revitalization of the IIED that continued to redefine ecodevelopment in the 1970s and Chip Linder's founding of the Center for Our Common Future in 1987. In doing so, UNEP assured that at least one NGO shared its normative values and would support its policies and programs.

5.5 Conditions Impacting Influence

To say that NGOs achieved their goals and influenced the norms of sustainable development is to tell only a part of the story as this analysis has yet to present information about the explanatory factors that identify the conditions under which NGOs achieve their goals, and thus have influence. In theorizing about NGO influence, five factors emerged that could potentially impact NGOs' ability to achieve their goals and to exert influence upon the sustainable development discourse. These factors are (1) normative traits and characteristics, (2) NGO capabilities, (3) political opportunities, (4) player's expectations, and (5) structural rules of engagement. It is highly likely these factors are not separable; that is, more than one factor may be present at any given point in time. Further, these multiple factors may have opposite effects on outcomes – one factor may positively impact NGO goal attainment, while a second factor may simultaneously negatively impact NGO goal attainment. Key conclusions about how these factors shape NGO influence on the sustainable development discourse are presented below. The conclusions are phrased in terms of the nature of the relationship between goal attainment (and therefore influence) and the factor under examination, as a positive/negative analysis may not be possible. For example, political circumstances may help an NGO achieve a goal in one case, and conspire against NGOs in differing circumstances.

5.5.1 Normative Traits and Characteristics

This section examines the relationship between goal attainment and normative traits and characteristics, including NGOs' use of rational moral authority. The examination begins by assuming that a status quo exists, and that the status quo represents a norm.

Conclusion 1: NGO influence appears to be inversely proportional to the difference between the new norm and the status quo.

More specifically, new norms that radically deviate from the current status quo constrain NGO influence. Large gaps between old and new norms suggest significant changes in behavior for regulated entities. Consequently, NGOs become less likely to find political allies among formal decision-makers while simultaneously increasing resistance to the new norm. The increase in resistance may make it harder for the NGO to find allies supporting the changes and harder to keep the policies and procedures dictated by the normative change alive on the international environmental negotiating agenda. On one hand, NGOs' conceptualization of sustainable development has remained largely intact, although contested – a point in NGOs favor. On the other hand, the absence of hard law on the subject suggests that there is considerable resistance to sustainable development on behalf of states – a point against NGOs' political influence.

It cannot be said that in any situation, only two competing norms exist. In truth, any number of competing norms may arise as winners or losers, or incomplete discourses may emerge as a result. Greater competition between norms suggests that actors will have a more difficult time achieving consensus.

Conclusion 2: NGO influence is also inversely proportional to the number of competing norms.

In the late 1970s multiple norms existed representing differing opinions about the relationship between environment and development such as ecodevelopment, sustainable development, irrational use of resources, and environmental stewardship. NGOs supporting sustainable development seek to influence actors supporting these other norms and will likely have to devise different tactics to deal with each group.

Conclusion 3: NGOs have more difficulty influencing norms and principles than rules and regulations.

To change a principle or norm is to also change a rule and regulation. However, it does not follow that the converse is true – that changes in rules and regulations represent fundamental shifts in principles and norms. In short, actors may do the "right" thing for the "wrong" reason. Japan's stance on whaling exemplifies this shift. NGOs successfully influenced the Stockholm Action Plan to include the moratorium on whaling. Japan had the ability to block this proviso, yet did not object. That is not to say that Japan believed in the underlying norm – that people should not hunt whales for food.

5.5.2 NGO Capabilities

The second conditioning factor, NGOs' capabilities, represents NGOs' influence due to their leveraging of material resources such as specialized knowledge, financial resources, and political expertise. NGOs' ability to transform scarce resources, whether knowledge, funding, or public opinion into more favorable treatment explains one of the most significant NGO victories and is an underlying factor in NGOs' ability to achieve all of their goals.

Conclusion 4: The greater the NGOs' capabilities the more likely NGOs will be able to achieve their goals.

Environmental politics has never been solely about the physical characteristics of the environment, despite Northern insistence on this position, but also about quality of life and the creation of wealth, along with the maintenance of natural resources. While NGOs with scientific knowledge and expertise provided invaluable support to the UN secretariat, non-state actors with economic expertise moved sustainable development forward at critical points. The political blend of environment and development agendas did not occur within the nation-state system, but rather from a network of like-minded NGOs and individuals. This network slowly conceptualized sustainable development beginning in 1974 with the Cocoyoc Declaration and continued to build on this framework until its adoption in the Brundtland Commission report in 1987.

This pattern of mixing environmental characteristics and development economics alleviated political opposition to the addition of the environment on the international agenda in the 1970s most especially at Founex and Cocoyoc. It is not an overstatement to say that sustainable development owes more to the developmental experts working on this issue in the 1970s than to the environmental activists during the same time. When environmental activists pursued items such as the conservation of natural resources, developmental economists merged this item with their own agenda, adding that natural resources should be well-used in order to promote both economic growth and social justice.

5.5.3 Political Circumstances

The third conditioning factor deals with political opportunities and deals with the timing of NGOs' goal attainment. NGO capabilities and material resources do not automatically equate to NGO goal attainment and influence. NGO resources may remain unused if there is not a political opportunity to do so. Political opportunities differ from structural rules of participation. They represent an obstacle to be overcome separate from the formal rules and procedures of the UN system.

Conclusion 5: NGOs with close relationships with UN bureaucrats have more opportunities to achieve their goals.

For example, NGOs influenced UNEP by providing free expertise and by allowing UNEP to use the NGO's public support networks. The lack of appropriate funding for UNEP created a problem that NGOs' resources solved. As a result, UNEP carefully constructed a political bargain that traded political access to UNEP, in exchange for a proactive lobbying group in support of the environment and development agenda. It then embedded this political bargain into the structure of the international environmental system, and in the process, gave NGOs a permanent structural location.

At the same time, however, it would also be incorrect to state that UNEP co-opted the NGOs involved in the case study. There is no evidence to suggest that UNEP successfully directed individual NGOs' activities, agendas or ideologies. Indeed, for a great deal of this historical case study, particularly in the pre-Brundtland era, NGOs served as a think-tank for UNEP. There was one case where UNEP was involved in pushing an NGO into supporting sustainable development (IUCN and the World Conservation Strategy in 1980), but even this attempt was only partially successful. This case study more accurately demonstrates that the UNEP-NGO relationship in the pre-Brundtland era was one of collusion, as opposed to cooptation. Further, this collusion was neither sinister nor malignant in the sense of UNEP deliberately setting out to take control of its NGO constituency, but rather that UNEP was aware of its own limitations in the system, namely its inability to advance an agenda.

Not all attempts to bargain will end successfully. Returning to the example above, completion of a political bargain may have unintended consequences that leave the NGO worse off than when they started.

Conclusion 6: NGOs with close relationships to decision-makers may also be constrained by these relationships.

One example of this is IUCN's contractual relationship with UNEP to produce the 1980 World Conservation Strategy. The first draft of the World Conservation Strategy in 1978 rejected the concepts of "basic needs". It was only after ELCI, a third world NGO, successfully involved UNEP (and presumably the withdrawal of UNEP funding) that sustainable development made its way into a major international policy document. Even so, IUCN unsurprisingly emphasized environmental protection over economic growth, a frame alignment that remained intact for the next decade.

5.5.4 Player's Expectations

The fourth conditioning factor deals with players expectations of NGOs. Arguably, the most visible change of the role of NGOs over the course of the entire case study was not a change in tactics or in political philosophy, but rather other actors perceptions and expectations about the capabilities and resources that NGOs were willing and able to commit to the sustainable development discourse.

States' attitude towards NGOs during the Stockholm conference are adequately captured by the ECO headliner "NGO come back, then go home." States did not expect NGOs to offer relevant political opinions or potential solutions. This limitation constrained NGOs' influence drastically (Eco 1972).

Conclusion 7: NGOs are more likely to obtain their goals when NGO communication is received by an open-minded player.

By the end of the case study NGOs were viewed as a "constituent" of UNEP along with developing countries. UN bureaucrats within UNEP actively listened and acted upon NGO advice and requests.

In contrast, there is little indication of NGO-state interaction after the Stockholm conference with respect to continued articulation of the sustainable development discourse.

Conclusion 8: NGOs are more likely to have influence in the absence of a dominant state.

That is not to say that states had no role in the case study, but rather to say that more powerful Western states generally were not paying attention to UNEP, ecodevelopment, or sustainable development beyond ensuring these actors and ideals did not seriously challenge Western economic growth and stability. There is no basis of comparison to make a conclusion about the demise of a state relative to other actors and certainly no evidence of a decline in the state system.

Chapter 3 defined power as one actor's continuous ability to influence a different actor over time. Despite the lengthy case study presented above, there is no conclusive evidence to suggest this in fact occurred. A careful review of this analysis shows that individual NGOs influenced a variety of actors from the earliest phases of global environmental diplomacy. The case study does not provide evidence to suggest that a specific NGO continuously influenced the same actor for the duration of this case study. Consequently, there is no evidence of NGOs' power.

5.5.5 Structural Rules of Engagement

The fifth enabling factor deals with the structural rules of engagement at the UN. The case study suggests that NGOs' role could be limited by the formal structures and tradition of the UN, in particular the procedural requirements of the UN mega-conferences that has, in the past, strongly favored states. However, NGOs must accept the limits imposed by the UN. In this case study, NGOs were not willing to accept the past restrictions and in response, the Stockholm conference secretariat co-opted NGOs into a parallel conference in order to lower NGO criticism of the conference and its outcomes.

Conclusion 9: Limits on formal participation may constrain NGO influence during international environmental negotiations, but these restrictions disappear after the conference ends.

In this sense, my research confirms Betsill and Corell (2008:15) when they assess that NGOs have become innovators at circumventing procedural requirements designed to limit their engagement. They further note that NGOs are more likely to have influence when states take positive steps to incorporate NGOs into negotiations and this research seems to support this conclusion.

Further, this research suggests that states have an easier time limiting NGO involvement during times of environmental negotiations. It is interesting to note that a state does not appear to have made an effort to restrict NGO access to UNEP. Thus, NGOs able to form relationships to stay in close contact with decision-makers beyond these meetings promote the likelihood of NGOs having influence.

Conclusion 10: NGOs with insider access to UN decision-makers enables NGO goal attainment and influence.

States failure to challenge NGOs' consultative relationship with UNEP allowed NGOs to promote themselves into the structure of the UN system. Generally speaking, the more informal the rules of access of NGO participation, the more influence wielded by NGOs. For example, UN bureaucrats at UNEP in Nairobi assisted in the construction of the sustainable development discourse. After Stockholm, the NGOs that most drastically influenced the normative conceptualization of sustainable development had regular and routine contact with UNEP bureaucrats. Organizations that came to Stockholm to observe the conference gave way to NGOs with "insider" contacts – both old and new.

These ten individual conclusions about NGO influence do not occur separately or sequentially. Rather, they should be thought of as layers of an integrated whole, with one layer assuming ascendancy over the others depending upon case specific circumstances. Nor are these conclusions the sum total of all layers of NGO influence; they are, however, all layers of NGO influence in this case study. As such, they may serve as generic indications of NGO influence upon the broader discourses within global environmental politics, and perhaps, international relations more broadly.

5.6 NGOs and the Cold War

This research provides strong evidence against associating increases in NGOs political participation at the UN with the end of the Cold War in 1992. The entire case study occurs during the Cold War and scholarly inquiry conducted at the time, albeit small in number of publications, document that NGOs achieved these goals. Rather, this research suggests, but does not investigate, that the NGO phenomenon at the Rio Earth Summit in 1992 replicated NGO patterns of behavior at Stockholm 20 years earlier.

Consequently, the formal institutionalization of NGOs at UNEP also occurred prior to the Cold War. This is subtly different from Martens' (2005) finding of institutionalization of UN-NGO relationships as a result of the end of the Cold War. The primary difference is one of scale. Claiming the institutionalization of NGOs across the entire UN system occurred in the 1970s is beyond the scope of this study as it is narrowly focused on the historic roots of sustainable development at UNEP. This research does support the professionalism and bureaucratization of NGOs as a result of the close working relationship that emerged during the early 1970s.

5.7 Conclusions

This chapter answered the question, to what extent have NGOs influenced the sustainable development discourse. It concluded that NGOs provided the primary conceptualization of sustainable development and they have influenced international environmental politics by seeking to obtain six primary goals. NGOs have engaged in political activities since the beginning of the international environmental agenda in the 1970s and have remained consistently active and influential. Somewhat unexpectedly, this research notes one significant change with respect to NGOs that occurred during this time frame. This change involves other actors' perception of NGOs – from apolitical observers, to cautious lobbyists, to full partners in sustainable development. This improvement is probably due to the recognition of NGOs' skills in project implementation, a function that had been carefully obscured from view by the nature of the UNEP-NGO consultancy relationship. What is "new" about NGO influence is that the results of NGOs' past successes, or in other words, their influence, are more visible in the present that in the past.

References

Alger, C. (1999). Strengthening relations between NGOs and the UN system: Towards a research agenda. *Global Society: Journal of Interdisciplinary International Relations, 13*(4), 393–409.

Betsill, M., & Corell, E. (2008). *NGO diplomacy: The influence of nongovernmental organizations in international environmental negotiations*. Cambridge: The MIT Press.

Eco. (1972). *NGO come back, then to home*. Bound as Stockholm conference Eco (Vol. 1). Stockholm, June 14, 1972.

Feraru, A. T. (1974). Transnational political interests and the global environment. *International Organization, 28*(1), 31–60.

Giddens, A. (1984). *The constitution of society: Outline of the theory of structuration*. Berkeley: University of California Press.

Keck, M. E., & Sikkink, K. (1998). *Activists beyond borders: Advocacy networks in international politics*. Ithaca: Cornell University Press.

Korten, D. C. (1990). *Getting to the 21st century voluntary action and the global agenda*. West Hartford: Kumarian Press.

Kütting, G., & Lipschutz, R. (2009). *Environmental governance: Power and knowledge in a local-global world*. London: Routledge.

Martens, K. (2005). *NGO's and the United Nations: Institutionalization, professionalization and adaptation*. Basingstoke: Palgrave Macmillan.

Princen, T., & Finger, M. (1994). *Environmental NGOs in world politics: Linking the local and the global*. London: Routledge.

Reimann, K. D. (2006). A view from the top: International politics, norms and the worldwide growth of NGOs. *International Studies Quarterly, 50*(1), 45–67.

Willetts, P. (1996). Consultative status for NGOs at the United Nations. In P. Willetts (Ed.), *The conscience of the world* (pp. 31–62). Washington, DC: Brookings Institution.

Chapter 6
Parting Thoughts

Abstract This chapter summarizes the key themes of this book – that nongovernmental organizations are an important, influential actor. NGOs political role, function and influence begins concurrently with the Stockholm conference and continues unabated throughout the 1970s and early 1980s. They successfully leveraged their rational moral authority into political influence that created a symbiotic relationship between United Nations Environment Program and nongovernmental organizations that fostered the sustainable development discourse. The chapter speculates that NGOs do not need sustainable development to harden into a formal regime in order to convince other actors to adopt its normative principles. In short, NGOs continue to work promoting the sustainable development discourse both within and outside the United Nations system. The chapter concludes by commenting that further research is needed to NGOs' political influence over extended periods of time, separate from specific international environmental negotiations.

Keywords Goal attainment • Nongovernmental Organizations (NGOs) • Influence • Conditioning factors • Rational moral authority • Cold War • Global environmental governance

6.1 Introduction

In the desire to conserve, manage, and protect the environment, questions about the relationship between human environment and development by necessity involve politics. Indeed, even the most scientific and "apolitical" statements about the quality of environmental protection have embedded within them a normative claim that it is good and proper to be concerned with the environment. As Boardman (1981) wrote "For the conservationist to argue that nature is apolitical can be a useful strategy. For him to actually believe this is a recipe for ineffectiveness." The prior chapters examined how conservationists, along with other environmentalists, worked out a political compromise, called sustainable development, to deal with

A.E. Egelston, *Sustainable Development*, DOI 10.1007/978-94-007-4878-1_6,

the relationships between human development and the environment. This chapter summarizes the main themes of this work and points towards future avenues of research.

6.2 Lessons Learned

NGOs arose to prominence within academic theory in recent years as scholars increasingly looked at these actors to see how they interacted with states at the international level. However, the historic role, functions and political influence of NGOs remains contested as scholars have difficulty agreeing upon who these groups are, what role and function they perform and whether or not these organizations influence the ordering of international affairs. This research seeks to contribute to this area by looking at these topics.

The study of sustainable development is complex in that it has an unparalleled number of actors engaged in a myriad of competing activities and a substantial negotiation process with no clear path forward. Although, considerable progress has been achieved over the last 30-years. The historical narrative begins with the introduction of a resolution at the UN to host the Conference on the Human Environment in 1968 and follows a series of actors and events forward in time to the beginnings of the Brundtland Commission in 1982.

NGOs political role, function, and influence begins concurrently with the Stockholm conference and continues unabated during this time frame. It was determined that NGOs obtain influence by achieving six key goals. These six key goals are getting information, providing expert assistance, lobbying decision-makers, representing the organization, implementing policy, and keeping the agenda alive. The inclusion of policy implementation adds to scholarly literature that in the past focused on NGOs contribution to the agenda-setting stage of regimes and regime formation. Because sustainable development is not a regime, theories about NGOs should give way to a broader spectrum of political activities that promote normative values and concrete actions. Policy implementation can no longer be viewed as a role solely for states.

Representing the organization and implementing policy are understudied and undertheorized, albeit for differing reasons. Representing the organization may be understudied due to the general nature of this goal and the significant lengths of time that may occur between action and result. Scholars have begun to research NGOs implementation of policy as part of a public-private partnership. However, observer bias continues to exist, and scholars seem to have difficulty moving past the expectation that NGOs only provide expert assistance and public support.

NGOs influence may be explained based upon five factors, or more precisely, combinations of these factors. These factors are (1) normative traits and characteristics, (2) NGOs' capabilities, (3) political circumstances, (4) actors' expectations, and (5) structural rules of engagement. NGOs have political influence both within and outside international environmental negotiations. NGOs generally support facets of the sustainable development discourse, but it would be misleading to conclude that NGOs uniformly support all aspects of sustainability.

Despite the determination that NGOs influence global environmental politics, declaring that NGOs are powerful actors in this arena seems premature. NGOs influence over environmental diplomacy is present, yet nevertheless inconsistent.

The evidence presented by this research contradicts scholarly theories that link the rise of NGOs to the end of the Cold War. While there has undoubtedly been growth in numbers of NGOs due to the end of the Cold War as well as the beginnings of greater political freedom behind the iron curtain, NGOs' influence within international environmental affairs is much older. It can, in fact, be substantiated and dated back to the early 1970s.

6.3 Implications

NGOs were a necessary actor in the historical conceptualization of sustainable development. Perhaps most intriguing about the research presented is the absence of the state in the spread of the norms underlying sustainable development. That is not to say that states are not an essential part of the structure of international society, but rather to point out that historically speaking, states did not conceptualize sustainable development. However, a seismic shift has occurred and state bureaucracies, particularly in Europe, have incorporated sustainable development into the highest levels of their government. Additionally, sustainable development benefits from growing support in North America. As states were not among the earlier supporters of sustainable development, it is nonsensical to speak of a hegemon enforcing sustainable development.

UNEP's willingness to engage every actor, through the principal of participation, has important implications for the direction of sustainable development. Widespread public participation allows for the partial bypassing of the international state system, by directing action towards other non-state actors that engage in environmentally destructive behavior – namely business and industry. A network of NGOs grouped around UNEP and G-77 allowed for the bypassing of the traditional international state system and the need for consensus before policy implementation. Concurrently, this capability to partially bypass the once sacrosanct state structure also allows success to be determined by something less than the formal declaration of international law.[1] Thus, every action taken on behalf of sustainable development, whether a symbolic trial program, a report analyzing the success and failures of sectors of business and industry, or a formal state program detailed through a national strategy, contributes to the spread and implementation of sustainable development.

[1] Sands (1989) distinguishes between international law as the law of and between states and that of international society, which has no formally recognized law.

The direct targeting of business and industry by proponents of sustainable development avoids the pitfalls of states inactivity due to the jealous guarding of their national sovereignty. That is, proponents of sustainable development do not need to convince a recalcitrant state to commit to changes in the legal structure to promote sustainable development. This can be particularly advantageous in the United States, where the likelihood of Congress passing legislation to implement sustainable development is unlikely, but whose businesses and industries are key components for the long-term viability of sustainable development.

The downside of bypassing states in the formation of sustainable development includes loss of the force of national law. Instead of convincing one state that there ought to be a law implementing the goals of sustainable development, each individual entity within the state must be convinced of the necessity of change. Additionally, there are no definitive penalties for noncompliance with sustainable development.

The reality of sustainable development falls somewhere between these two extremes. In the case of Western Europe, countries such as Norway, Sweden and the Netherlands have wholeheartedly embraced the rhetoric and the activities necessary for implementing sustainable development. Proponents of sustainable development have welcomed these countries' activities and have actively encouraged their continued implementation and support. In "laggard" countries such as Australia and the United States, pro-environmental groups have found it necessary to bypass the formal legislative system and to negotiate directly with friendly companies such as DuPont, who have reaped substantial financial rewards for their willingness to participate in the sustainable development discourse.

6.4 One Way Forward

During the process of writing this book, climate change has moved onto the list of public policy measures that ripened for action in the United States, thanks in no small measure, to Al Gore's Oscar-winning documentary *An Inconceivable Truth*. Americans are also paying more attention to fuel efficiency in cars along with other energy saving measures, albeit more out of attention to the ongoing crisis in the Middle East than any environmental leanings. Proponents of sustainable development nevertheless have an opportunity to tap into these current events in order to promote a healthier environment and a more just division of goods and services across the world. If such action is to be forthcoming, it will not happen because of the United States government, but rather because ordinary citizens and their action groups, NGOs, influenced others to change behaviors.

Despite growing support among the general population, congressional action on climate change remains uncertain, and any kind of legislation involving sustainable development is highly unlikely. Yet it is not clear that congressional action is necessary to spread sustainable development. Because the system being altered is the Western socio-economic structure, the possibility for change separate from state intervention exists. As Lipschutz (1996) theorized, and this research confirms, changes in rules and /or systems of rules away from states is an area in which NGOs excel. Historically speaking, sustainable development has been significantly shaped by the political influence, and rational moral authority of these non-state actors. Further, NGOs have not significantly altered their political sources of rational moral authority in spearheading sustainable development since its origins in the 1970s.

It is therefore not surprising to see within sustainable development various types of non-state actors providing a level of global environmental governance through their activities of knowledge provider, normative developer, political lobbyist, or project implementer. One fundamentally important question about sustainable development remains, however. How cohesive is this system, or in the words of Lipschutz (1996:238) "Does order or chaos prevail?"

The differing intellectual histories of sustainable development undoubtedly contribute to the chaos element within sustainable development. Sustainable development continues to represent significantly differing norms. These differing norms in turn lead to differing political agendas over economic growth, international development aid, and consumption of finite resources at both the manufacturing and personal levels. It is not surprising that diplomats have been unable to agree to the formation of an international regime.

Despite the fact that at least three distinctive discourses on sustainable development exist (limits discourse, growth discourse and resource conservation) the actions generally required by the three discourses are remarkable similar. To be sure, a plurality of rule-setters increases the probability of different rules, along with deliberate manipulation of the rules for individual group purposes, but the formation of networks has the effect of converging standards of behavior into what could be characterized as predictable chaos.

6.5 Directions for Further Research

The most important direction for future research would be the confirmation of NGOs' political influence over extended periods of time, separate from specific international environmental negotiations. The argument that the process of international governance is a continuous process has considerable merit. Following this logic, the lack of a definitive end to the process does not necessarily mean that theorizing about the process of international governance is untenable.

Examining the process of international environmental governance seems to require a case study with a well-documented history, a complexity of actors, and ideally, widespread implementation. Case studies that might meet these requirements include climate change and biodiversity. Additionally, the relative importance of NGOs might be further reviewed through examination of a case where NGO activity would seem likely, but is, in fact, absent. The Convention on Chemical Weaponry would seem ideal as one can construct an argument that some of the same NGOs that are active in this case study, namely ENGOs, in theory have the correct technical expertise to meaningfully contribute to this regime.

One important avenue of research that could stem from this case study includes a comparative analysis of the Stockholm to Rio process versus Rio to Johannesburg, or even Rio '92 to Rio '12. The undertaking to do so would be complex and detailed, as the NGO block has expanded considerably since 1972. This expansion also denotes the continued NGO fragmentation around the sustainable development discourse.

6.6 Conclusions

To summarize, the analysis of sustainable development suggests that the process of international governance is more complicated than a simple analysis of the state-led negotiating process as it incorporates NGOs as actors with significant political influence over the agenda, ideology, and implementation of sustainable development. These non-state actors' influence is generally much older than academic theory suggests as the political influence of these groups have their origins in the 1970s. The recognition of their political influence suggests that mandatory regulation through the use of international law is one pathway towards change in behavior. However, it is not the only road as alternatives to formal regulations are now being explored concurrently with the lack of substantial progress at the UN on the implementation of sustainable development. So we wait and hope with great expectations, tempered with the knowledge of past successes and failures, that NGOs attempts to move sustainability forward would succeed in the protection of our planet.

References

Boardman, R. (1981). *International organization and the conservation of nature*. Bloomington: Indiana University Press.

Lipschutz, R. D. (1996). *Global civil society and global environmental governance: The politics of nature from place to planet*. Albany: State University of New York Press.

Sands, P. (1989). The environment, community and international law. *Harvard International Law Journal, 30*(2), 393–420.

Index

A.E. Egelston, *Sustainable Development*, DOI 10.1007/978-94-007-4878-1,

MIX
Papier aus verantwortungsvollen Quellen
Paper from responsible sources
FSC® C105338

If you have any concerns about our products,
you can contact us on
ProductSafety@springernature.com

In case Publisher is established outside the EU,
the EU authorized representative is:
Springer Nature Customer Service Center GmbH
Europaplatz 3, 69115 Heidelberg, Germany

Printed by Libri Plureos GmbH
in Hamburg, Germany